T0201500

*How To Prepare Defense-Related Scientific and Technical Reports*

## THE WILEY BICENTENNIAL–KNOWLEDGE FOR GENERATIONS

*E*ach generation has its unique needs and aspirations. When Charles Wiley first opened his small printing shop in lower Manhattan in 1807, it was a generation of boundless potential searching for an identity. And we were there, helping to define a new American literary tradition. Over half a century later, in the midst of the Second Industrial Revolution, it was a generation focused on building the future. Once again, we were there, supplying the critical scientific, technical, and engineering knowledge that helped frame the world. Throughout the 20th Century, and into the new millennium, nations began to reach out beyond their own borders and a new international community was born. Wiley was there, expanding its operations around the world to enable a global exchange of ideas, opinions, and know-how.

For 200 years, Wiley has been an integral part of each generation's journey, enabling the flow of information and understanding necessary to meet their needs and fulfill their aspirations. Today, bold new technologies are changing the way we live and learn. Wiley will be there, providing you the must-have knowledge you need to imagine new worlds, new possibilities, and new opportunities.

Generations come and go, but you can always count on Wiley to provide you the knowledge you need, when and where you need it!

**WILLIAM J. PESCE**
PRESIDENT AND CHIEF EXECUTIVE OFFICER

**PETER BOOTH WILEY**
CHAIRMAN OF THE BOARD

# How To Prepare Defense-Related Scientific and Technical Reports

## Guidance for Government, Academia, and Industry

**Walter W. Rice**

WILEY-INTERSCIENCE
A JOHN WILEY & SONS, INC., PUBLICATION

Published by John Wiley & Sons, Inc., Hoboken, New Jersey.
Published simultaneously in Canada.

For general information on our other products and services or for technical support, please contact our
Customer Care Department within the United States at (800) 762-2974, outside the United States at
(317) 572-3993 or fax (317) 572-4002.

Wiley also publishes its books in a variety of electronic formats. Some content that appears in print may
not be available in electronic format. For information about Wiley products, visit our web site at
www.wiley.com.

*Library of Congress Cataloging-in-Publication Data is available.*

ISBN-13: 978-0-471-72509-1
ISBN-10: 0-471-72509-9

Printed in the United States of America.

10 9 8 7 6 5 4 3 2 1

To my wife, Myra, and my sons, Kevin, Brian, and Darren

# Contents

|  | *Page* |
|---|---|
| **Figures** | xiii |
| **Tables** | xix |
| **Foreword** | xxi |
| **Preface** | xxiii |
| Purpose and Audience | xxiii |
| Need for Standardization | xxiv |
| Creation, Distribution, and Preservation of Scientific and Technical Information | xxv |
| Approach | xxv |
| **Acknowledgments** | xxix |
| **Chapter 1: Background** | 1 |
| 1.1 Definition of a Technical Report | 1 |
| 1.2 Standards | 4 |
| 1.2.1 History | 4 |
| 1.2.2 ANSI/NISO Z39.18-2005, "Scientific and Technical Reports - Preparation, Presentation and Preservation" | 5 |
| 1.2.3 Related NISO Standards | 7 |
| 1.2.3.1 ANSI/NISO Z39.23-1997, "Standard Technical Report Number Format and Creation" | 7 |

*Page*

1.2.3.2 ANSI/NISO Z39.14-1997, "Guidelines for
Abstracts" ...................................... 7
1.3 Department of Defense's Scientific and Technical
Information Program ................................ 8
1.4 Publishing Requirements ........................... 8
1.4.1 Author Responsibilities .......................... 9
1.4.2 Non-Author Responsibilities ..................... 10
1.5 Review of Technical Reports ....................... 10
1.5.1 Technical Review ............................... 11
1.5.2 Security Review ................................ 11
1.5.3 Distribution Statement Review ................... 12
1.5.4 Export Control Review .......................... 14
1.5.5 Patent and Technology Transfer Review ........... 14
1.6 Distribution of Technical Reports .................. 15
1.6.1 Basic Guidelines .............................. 15
1.6.2 Primary Distribution ........................... 15
1.6.2.1 Internal Distribution ......................... 16
1.6.2.2 External Distribution ......................... 16
1.6.2.2.1 General Guidelines ....................... 16
1.6.2.2.2 Secondary Distribution ................... 17
1.6.2.2.3 Foreign Distribution ..................... 17
1.7 Classified Technical Reports ....................... 20
1.7.1 Description .................................... 20
1.7.2 Author Responsibilities ......................... 21
1.7.3 Original and Derivative Classification ............. 22
1.7.4 Marking Requirements for Classified Technical
Reports ......................................... 22
**Chapter 2: Organization and Design** ................... 23
2.1 Organization ...................................... 23
2.2 Design ........................................... 25
2.2.1 Subordination and Format ....................... 25
2.2.1.1 Classification Markings ....................... 25
2.2.1.2 Typographical Progression .................... 27
2.2.1.2.1 Headings .............................. 27
2.2.1.2.2 Paragraphs ............................ 28
2.2.1.2.3 Lists .................................. 30
2.2.1.3 Decimal Numbering .......................... 31
2.2.1.3.1 Headings .............................. 31
2.2.1.3.2 Paragraphs ............................ 34
2.2.1.3.3 Lists .................................. 34

*Page*

2.2.2 Visuals . . . . . . . . . . . . . . . . . . . . . . . . . . . . . . . . . . .   34
  2.2.2.1 Description . . . . . . . . . . . . . . . . . . . . . . . . . . . . .   34
  2.2.2.2 Numbering and Titling . . . . . . . . . . . . . . . . . . . . .   37
  2.2.2.3 Classified Technical Reports . . . . . . . . . . . . . . . .   38
  2.2.2.4 Examples . . . . . . . . . . . . . . . . . . . . . . . . . . . . . .   38
  2.2.2.5 Figures . . . . . . . . . . . . . . . . . . . . . . . . . . . . . . .   41
  2.2.2.6 Tables . . . . . . . . . . . . . . . . . . . . . . . . . . . . . . . .   49
2.2.3 Page Format . . . . . . . . . . . . . . . . . . . . . . . . . . . . . . . .   50
  2.2.3.1 Column Format and Line Length . . . . . . . . . . . . .   50
  2.2.3.2 Typography . . . . . . . . . . . . . . . . . . . . . . . . . . . .   50
  2.2.3.3 Image Area . . . . . . . . . . . . . . . . . . . . . . . . . . . .   52
  2.2.3.4 Margins . . . . . . . . . . . . . . . . . . . . . . . . . . . . . .   53
  2.2.3.5 Page Numbers . . . . . . . . . . . . . . . . . . . . . . . . . .   53
  2.2.3.6 Printers and Paper and Ink . . . . . . . . . . . . . . . . .   56
  2.2.3.7 Compilation of Information in Classified
    Technical Reports . . . . . . . . . . . . . . . . . . . . . . . . .   56
    2.2.3.7.1 Unclassified Portions Requiring Overall
      Classification . . . . . . . . . . . . . . . . . . . . . . . . . . .   56
    2.2.3.7.2 Classified Portions Requiring Higher
      Overall Classification . . . . . . . . . . . . . . . . . . . . .   57
  2.2.3.8 Blank Pages . . . . . . . . . . . . . . . . . . . . . . . . . . . .   57
2.2.4 Equations . . . . . . . . . . . . . . . . . . . . . . . . . . . . . . . . . .   57
2.2.5 Footnotes . . . . . . . . . . . . . . . . . . . . . . . . . . . . . . . . . .   61
2.2.6 Errata . . . . . . . . . . . . . . . . . . . . . . . . . . . . . . . . . . . . .   61
**Chapter 3: Front Matter** . . . . . . . . . . . . . . . . . . . . . . . . . . . . .   65
3.1 Front Cover and Title Page . . . . . . . . . . . . . . . . . . . . . . . .   65
3.2 Inside Front Cover . . . . . . . . . . . . . . . . . . . . . . . . . . . . . .   67
3.3 Back Cover . . . . . . . . . . . . . . . . . . . . . . . . . . . . . . . . . . .   68
3.4 Back of Title Page . . . . . . . . . . . . . . . . . . . . . . . . . . . . . .   68
3.5 Technical Report Number . . . . . . . . . . . . . . . . . . . . . . . . .   69
3.6 Document Control Number . . . . . . . . . . . . . . . . . . . . . . . .   71
3.7 Performing Organization and Sponsoring Organization . . .   72
3.8 Publication Date . . . . . . . . . . . . . . . . . . . . . . . . . . . . . . . .   72
3.9 Title and Subtitle . . . . . . . . . . . . . . . . . . . . . . . . . . . . . . .   72
3.10 Titling and Numbering of Series . . . . . . . . . . . . . . . . . . .   73
3.11 Authorship . . . . . . . . . . . . . . . . . . . . . . . . . . . . . . . . . . .   73
3.12 Notices . . . . . . . . . . . . . . . . . . . . . . . . . . . . . . . . . . . . . .   76
  3.12.1 Overall Classification Level . . . . . . . . . . . . . . . . . . .   76
  3.12.2 Unclassified Controlled Information . . . . . . . . . . . . .   76
    3.12.2.1 For Official Use Only (FOUO) . . . . . . . . . . . . . .   77

*Page*

3.12.2.2 Sensitive But Unclassified (SBU) Information ..    80
3.12.2.3 Drug Enforcement Administration (DEA)
Sensitive Information ............................    80
3.12.2.4 DoD Unclassified Controlled Nuclear
Information (DoD UCNI) ........................    81
3.12.2.5 Sensitive Information (Computer Security Act
of 1987) .........................................    81
3.12.3 Intelligence Control Markings ..................    81
3.12.4 Warning Notices ..............................    82
3.12.5 Restrictive Markings on Noncommercial Data
and Software .......................................    84
3.12.6 Distribution Statements ........................    85
3.12.6.1 Distribution Statement Checklist .............    86
3.12.6.2 Definitions of Reasons for Limiting
Distribution of Technical Reports ..................    87
3.12.6.3 Date of Determination ......................    89
3.12.6.4 Further Requests to Controlling Office ........    89
3.12.6.5 Use of Distribution Statements ..............    89
3.12.6.5.1 Distribution Statement A ................    89
3.12.6.5.2 Distribution Statements B, C, D, and E ....    90
3.12.6.5.3 Distribution Statement F ................    90
3.12.6.5.4 Distribution Statement X ................    90
3.12.7 Export Control Notice .........................    91
3.12.8 Downgrade, Declassify, and Related Markings .....    93
3.12.9 Destruction Notice ............................    98
3.12.10 Foreign Government and North Atlantic Treaty
Organization (NATO) Information ....................    99
3.13 Seals/Emblems and Logos ..........................    101
3.14 Example Covers and Title Pages .....................    101
3.15 Report Documentation Page (Standard Form 298) ......    102
3.16 Abstract .........................................    117
3.17 Table of Contents .................................    119
3.18 List(s) of Figures and Tables .......................    125
3.19 Foreword ........................................    133
3.20 Preface (Administrative Information) .................    137
3.21 Acknowledgments .................................    140
**Chapter 4: Body** ......................................    145
4.1 Executive Summary ................................    146
4.2 Introduction ......................................    147
4.3 Methods, Assumptions, and Procedures ................    156

*Page*

4.4 Results and Discussion ............................. 157
4.5 Conclusions ...................................... 157
4.6 Recommendations ................................. 158
4.7 Examples of Main Body ........................... 158
4.8 List of References ................................ 161
**Chapter 5: Back Matter** ............................ 187
5.1 Appendixes ...................................... 187
5.2 Bibliography ..................................... 192
5.3 List(s) of Symbols, Abbreviations, and Acronyms ....... 205
5.4 Glossary ........................................ 209
5.5 Index ........................................... 213
5.6 Distribution List ................................. 218
5.7 Back Cover ...................................... 223
**References** ........................................ 229
**Appendix A: Defense Technical Information Center**
            **(DTIC)** ................................ 233
A.1 Overview ........................................ 233
A.2 DTIC Databases ................................. 237
   A.2.1 Technical Report (TR) Database .................. 238
      A.2.1.1 Overview ............................... 238
      A.2.1.2 Submission Guidelines .................... 238
   A.2.2 Research Summaries (RS) Database ............... 239
   A.2.3 Independent Research and Development (IR&D)
         Database .................................... 239
   A.2.4 Research and Development Descriptive
         Summaries (RDDS) Database .................... 240
A.3 Information Analysis Centers (IACs) ................. 240
A.4 Registration with DTIC .......................... 242
A.5 Research at DTIC ................................ 243
   A.5.1 Scientific and Technical Information Network
         (STINET) ................................... 243
      A.5.1.1 Public STINET .......................... 243
      A.5.1.2 Private STINET ......................... 243
      A.5.1.3 Classified STINET ....................... 244
   A.5.2 Handle Service .............................. 244
   A.5.3 QuestionPoint ("Ask a Librarian") .............. 244
**Appendix B: Tone and Style** ........................ 245
B.1 Tone ........................................... 245
B.2 Style .......................................... 248

*Page*

B.2.1 United States Government Printing Office Style
Manual ............................................. 248
B.2.1.1 Editor's and Illustrator's Marks ............... 249
B.2.1.2 Capitalization ............................... 249
B.2.1.2.1 Derivatives of Proper Names ............. 249
B.2.1.2.2 Names of Military Organizations .......... 249
B.2.1.2.3 Scientific Names ........................ 249
B.2.1.2.4 Titles of Persons ........................ 251
B.2.1.2.5 Capitalization and Spelling Examples ...... 251
B.2.1.3 Compounding Rules ......................... 264
B.2.1.3.1 Military Titles ........................... 264
B.2.1.3.2 Scientific and Technical Terms ............ 264
B.2.1.4 Symbols, Abbreviations, and Acronyms ........ 265
B.2.1.4.1 Symbols and Abbreviations ............... 265
B.2.1.4.2 Acronyms ............................... 265
B.2.1.4.3 Standard Abbreviations .................. 267
B.2.1.4.4 Terms of Measure ....................... 285
B.2.1.4.5 Units of Measurement ................... 286
B.2.1.4.6 Chemical Names and Abbreviations ........ 298
B.2.1.5 Italic ...................................... 300
B.2.1.5.1 Names of Aircraft, Vessels, and
Spacecraft ................................... 300
B.2.1.5.2 Scientific Names ........................ 301
B.2.1.5.3 Letter Designations ..................... 301
B.2.1.5.4 Numerals ............................... 301
B.2.2 The Chicago Manual of Style ................... 304
B.2.3 Other Style Manuals ........................... 305
**Bibliography** ......................................... 307
Reference Literature ................................... 307
Technical Communication Literature .................... 308
**Abbreviations** ....................................... 315
**Glossary** ............................................ 319
**Index** ............................................... 327

# *Figures*

| | | *Page* |
|---|---|---|
| 1-1 | Warning, Caution, and Note . . . . . . . . . . . . . . . . . . . . . . . . . . | 3 |
| 1-2 | DTIC Form 55 (Page 1 of 2) . . . . . . . . . . . . . . . . . . . . . . . . | 18 |
| 1-3 | DTIC Form 55 (Page 2 of 2) . . . . . . . . . . . . . . . . . . . . . . . . | 19 |
| 2-1 | Examples of Headings Using Typographical Progression Format . . . . . . . . . . . . . . . . . . . . . . . . . . . . . . . . . . . . . . . . | 29 |
| 2-2 | Examples of Lists Using Typographical Progression Format . . . . . . . . . . . . . . . . . . . . . . . . . . . . . . . . . . . . . . . . | 32 |
| 2-3 | Examples of Headings Using Decimal Numbering Format . . . . . . . . . . . . . . . . . . . . . . . . . . . . . . . . . . . . . . . . | 33 |
| 2-4 | Examples of Lists Using Decimal Numbering Format . . . . | 35 |
| 2-5 | Examples of Portrait and Landscape Orientations in Unclassified Technical Report . . . . . . . . . . . . . . . . . . . . . . . | 39 |
| 2-6 | Examples of Portrait and Landscape Orientations in Classified Technical Report . . . . . . . . . . . . . . . . . . . . . . . . | 40 |
| 2-7 | Examples of Subfigures and Facing Pages in Unclassified Technical Report . . . . . . . . . . . . . . . . . . . . . . . . . . . . . . . . | 42 |
| 2-8 | Examples of Subfigures and Facing Pages in Classified Technical Report . . . . . . . . . . . . . . . . . . . . . . . . . . . . . . . . | 43 |
| 2-9 | Example of 17-Inch-by-11-Inch Foldout in Unclassified Technical Report . . . . . . . . . . . . . . . . . . . . . . . . . . . . . . . . | 44 |

*Page*

2-10  Example of 22-Inch-by-11-Inch Foldout in Unclassified
      Technical Report ................................    44
2-11  Example of 17-Inch-by-11-Inch Foldout in Classified
      Technical Report ................................    45
2-12  Example of 22-Inch-by-11-Inch Foldout in Classified
      Technical Report ................................    45
2-13  Information Absorption ...........................    46
2-14  Information Retention ............................    46
2-15  Figure Components and Classification Markings .........    48
2-16  Table Components and Classification Markings .........    51
2-17  Image Areas and Margins in Unclassified Technical
      Report ..........................................    54
2-18  Image Areas and Margins in Classified Technical
      Report ..........................................    55
2-19  Examples of Equations in Unclassified and Classified
      Technical Reports ................................    60
2-20  Examples of Footnotes in Unclassified and Classified
      Technical Reports ................................    62
2-21  Errata Sheet Format ..............................    63
3-1   Example of Unclassified Front Cover ................    103
3-2   Example of Unclassified Title Page ..................    104
3-3   Example of Classified Front Cover ..................    105
3-4   Example of Classified Title Page ...................    106
3-5   Examples of Inside Front Cover ....................    107
3-6   Example of Back of Title Page (Unclassified Technical
      Report) .........................................    108
3-7   Example of Back of Title Page (Classified Technical
      Report) .........................................    109
3-8   Example of Back Cover of Unclassified Technical
      Report ..........................................    110
3-9   Example of Back Cover of Classified Technical Report ...    111
3-10  Standard Form 298 (Front) .........................    112
3-11  Standard Form 298 (Back) .........................    113
3-12  Example of Abstract in Unclassified Technical Report ....    120
3-13  Example of Abstract in Classified Technical Report ......    121
3-14  Headings at Same Level ...........................    122
3-15  Minimum Number of Headings ......................    124
3-16  Example of Table of Contents in Unclassified
      Technical Report Using Typographical Progression
      Format ..........................................    126

*Page*

3-17  Example of Unclassified Table of Contents in Classified
Technical Report Using Typographical Progression
Format . . . . . . . . . . . . . . . . . . . . . . . . . . . . . . . . . . . . . . . .    127
3-18  Example of Classified Table of Contents in Classified
Technical Report Using Typographical Progression
Format . . . . . . . . . . . . . . . . . . . . . . . . . . . . . . . . . . . . . . . .    128
3-19  Example of Table of Contents in Unclassified Technical
Report Using Decimal Numbering Format . . . . . . . . . . . . .    129
3-20  Example of Unclassified Table of Contents in Classified
Technical Report Using Decimal Numbering Format . . . . .    130
3-21  Example of Classified Table of Contents in Classified
Technical Report Using Decimal Numbering Format . . . . .    131
3-22  Example of Lists of Figures and Tables in Unclassified
Technical Report Using Typographical Progression
Format . . . . . . . . . . . . . . . . . . . . . . . . . . . . . . . . . . . . . . . .    134
3-23  Example of Unclassified Lists of Figures and Tables in
Classified Technical Report Using Typographical
Progression Format . . . . . . . . . . . . . . . . . . . . . . . . . . . . . . .    135
3-24  Example of Classified Lists of Figures and Tables in
Classified Technical Report Using Typographical
Progression Format . . . . . . . . . . . . . . . . . . . . . . . . . . . . . . .    136
3-25  Example of Lists of Figures and Tables in Unclassified
Technical Report Using Decimal Numbering Format . . . . .    137
3-26  Example of Unclassified Lists of Figures and Tables in
Classified Technical Report Using Decimal Numbering
Format . . . . . . . . . . . . . . . . . . . . . . . . . . . . . . . . . . . . . . . .    138
3-27  Example of Classified Lists of Figures and Tables in
Classified Technical Report Using Decimal Numbering
Format . . . . . . . . . . . . . . . . . . . . . . . . . . . . . . . . . . . . . . . .    139
3-28  Example of Foreword, Preface (Administrative
Information), or Acknowledgments in Unclassified
Technical Report . . . . . . . . . . . . . . . . . . . . . . . . . . . . . . . . .    142
3-29  Example of Foreword, Preface (Administrative
Information), or Acknowledgments in Classified
Technical Report . . . . . . . . . . . . . . . . . . . . . . . . . . . . . . . . .    143
4-1   Example of Executive Summary in Unclassified
Technical Report Using Typographical Progression
Format . . . . . . . . . . . . . . . . . . . . . . . . . . . . . . . . . . . . . . . .    148

*Page*

4-2   Example of Executive Summary in Classified
      Technical Report Using Typographical Progression
      Format . . . . . . . . . . . . . . . . . . . . . . . . . . . . . . . . . . . . . . . . . .    150
4-3   Example of Executive Summary in Unclassified Technical
      Report Using Decimal Numbering Format . . . . . . . . . . . . . .    152
4-4   Example of Executive Summary in Classified Technical
      Report Using Decimal Numbering Format . . . . . . . . . . . . . .    154
4-5   Example of Main Body in Unclassified Technical
      Report Using Typographical Progression Format . . . . . . . .    159
4-6   Example of Main Body in Classified Technical
      Report Using Typographical Progression Format . . . . . . . .    163
4-7   Example of Main Body in Unclassified Technical
      Report Using Decimal Numbering Format . . . . . . . . . . . . . .    167
4-8   Example of Main Body in Classified Technical
      Report Using Decimal Numbering Format . . . . . . . . . . . . . .    171
4-9   Example List of References in Unclassified Technical
      Report Using Typographical Progression Format . . . . . . . .    177
4-10  Example List of References in Classified Technical
      Report Using Typographical Progression Format . . . . . . . .    178
4-11  Example List of References in Unclassified Technical
      Report Using Decimal Numbering Format . . . . . . . . . . . . . .    179
4-12  Example List of References in Classified Technical
      Report Using Decimal Numbering Format . . . . . . . . . . . . . .    180
5-1   Example of Appendix in Unclassified Technical
      Report Using Typographical Progression Format . . . . . . . .    190
5-2   Example of Cover Sheet of Unclassified Appendix in
      Classified Technical Report Using Typographical
      Progression Format . . . . . . . . . . . . . . . . . . . . . . . . . . . . . . . .    192
5-3   Example of Body of Unclassified Appendix in Classified
      Technical Report Using Typographical Progression
      Format . . . . . . . . . . . . . . . . . . . . . . . . . . . . . . . . . . . . . . . . . .    193
5-4   Example of Classified Appendix in Classified Technical
      Report Using Typographical Progression Format . . . . . . . .    195
5-5   Example of Appendix in Unclassified Technical
      Report Using Decimal Numbering Format . . . . . . . . . . . . . .    197
5-6   Example of Cover Sheet of Unclassified Appendix in
      Classified Technical Report Using Decimal Numbering
      Format . . . . . . . . . . . . . . . . . . . . . . . . . . . . . . . . . . . . . . . . . .    199
5-7   Example of Body of Unclassified Appendix in Classified
      Technical Report Using Decimal Numbering Format . . . . .    200

*Page*

5-8   Example of Classified Appendix in Classified Technical
        Report Using Decimal Numbering Format .............   202
5-9   Example of Bibliography in Unclassified Technical
        Report Using Typographical Progression Format ........   204
5-10  Example of Bibliography in Classified Technical Report
        Using Typographical Progression Format ..............   205
5-11  Example of Bibliography in Unclassified Technical Report
        Using Decimal Numbering Format ...................   206
5-12  Example of Bibliography in Classified Technical Report
        Using Decimal Numbering Format ...................   207
5-13  Example of List of Symbols, Abbreviations, and
        Acronyms in Unclassified Technical Report Using
        Typographical Progression Format ..................   210
5-14  Example of List of Symbols, Abbreviations, and
        Acronyms in Classified Technical Report Using
        Typographical Progression Format ..................   211
5-15  Example of List of Symbols, Abbreviations, and
        Acronyms in Unclassified Technical Report Using
        Decimal Numbering Format .........................   212
5-16  Example of List of Symbols, Abbreviations, and
        Acronyms in Classified Technical Report Using Decimal
        Numbering Format ................................   213
5-17  Example of Glossary in Unclassified Technical Report
        Using Typographical Progression Format ..............   214
5-18  Example of Glossary in Classified Technical Report
        Using Typographical Progression Format ..............   215
5-19  Example of Glossary in Unclassified Technical Report
        Using Decimal Numbering Format ...................   216
5-20  Example of Glossary in Classified Technical Report Using
        Decimal Numbering Format .........................   217
5-21  Example of Index in Unclassified Technical Report
        Using Typographical Progression Format ..............   219
5-22  Example of Index in Classified Technical Report
        Using Typographical Progression Format ..............   220
5-23  Example of Index in Unclassified Technical Report Using
        Decimal Numbering Format .........................   221
5-24  Example of Index in Classified Technical Report Using
        Decimal Numbering Format .........................   222
5-25  Example of Distribution List in Unclassified Technical
        Report Using Typographical Progression Format ........   224

*Page*

5-26  Example of Distribution List in Classified Technical
      Report Using Typographical Progression Format ........  225
5-27  Example of Distribution List in Unclassified Technical
      Report Using Decimal Numbering Format ..............  226
5-28  Example of Distribution List in Classified Technical
      Report Using Decimal Numbering Format ..............  227
A-1   DTIC Core Collections by Level of Access ............  237
B-1   Editor's and Illustrator's Marks ......................  250

# *Tables*

|  |  | Page |
|---|---|---|
| 2-1 | Technical Report Elements ......................... | 24 |
| 3-1 | Military Title Abbreviations ....................... | 75 |
| 3-2 | Distribution Statements ........................... | 85 |
| 3-3 | Distribution Statement Checklist ................... | 86 |
| 3-4 | Exemption Categories ............................ | 96 |
| 3-5 | Foreign Classification Levels ...................... | 99 |
| 3-6 | Foreign Country Abbreviations ..................... | 114 |
| 3-7 | Warning Notice Abbreviations ..................... | 115 |
| 4-1 | Example References .............................. | 181 |
| B-1 | Standard Abbreviations ........................... | 267 |
| B-2 | Standard Abbreviations and Symbols for Units of Measurement ...................................... | 286 |
| B-3 | Chemical Names and Abbreviations ................. | 298 |

# *Foreword*

In 1981 Krishna Subramanyam wrote that the quest for scientific knowledge is an evolutionary process in which every increment of new knowledge adds to, modifies, refines, or refutes earlier findings.

Few tasks are less desired and more intimidating than documenting the exciting scientific or technical work one has just completed. The fact remains, however, that scientific and technical information is a key element in scientific and technical knowledge advances, whether done as an individual effort or through collaboration.

Because science is the result of the accumulated knowledge of many, there is a clear need to document scientific and technical efforts, whether or not they have succeeded. In the Department of Defense (DoD) it is more than a professional obligation, it is normally a contractual obligation if a contractor performs the work. But how does one go about documenting work such as a technical report that is not covered by the publishing requirements of a journal publisher?

I was privileged to chair the National Information Standards Organization (NISO) Committee AW charged with updating ANSI/NISO Z39.18. In developing the new standard, the committee faced several challenges. It was tasked to update the version of the standard that was published in 1995. ANSI/NISO Z39.18-1995 was based entirely on "paper and ink" documentation. The revision, ANSI/NISO Z39.18-2005, "Scientific and Technical Reports - Preparation, Presentation, and Preservation," recognized that "paper and ink," although still a principal documentation base, is only one of several alternatives now available and that "paper and ink" may be used less in the future. The committee also recognized that this transition was not amenable to developing a strict standard. Therefore, the revised standard presents alternatives and guidelines that should be used based on report purpose and presentation.

As a former Defense Technical Information Center (DTIC) Administrator, I am pleased that this book includes an extensive appendix covering DTIC, DoD's central facility for scientific and technical information. DTIC serves as a vital link in the transfer of information among DoD personnel, DoD contractors and potential contractors, and other U.S. Government agency personnel and their contractors. The book also contains a very valuable appendix on tone and style, an area that always is a challenge. The object of technical documentation is to place information and knowledge into the brain of someone who will produce more knowledge. If the technical report writing and editing process does not produce a document that does not communicate well, it fails in its function.

Walter Rice, in his excellent book, takes a necessary next step by taking guidelines from the revised NISO standard and other standards and adding DoD requirements. Walter was on the committee that produced ANSI/NISO Z39.18-1995. This, plus his extensive knowledge based on his years of experience editing scientific and technical reports in the U.S. Government, industry, and, now, academia, makes his book a very valuable resource for all who create, manage, distribute, and archive defense-related technical reports.

It is one thing to be knowledgeable; it is another thing to be able to communicate this knowledge. Walter does this well, addressing in clear language and useful illustrations and tables everything needed by all the participants in the scientific and technical information continuum, from authors to librarians to archivists.

KURT MOLHOLM
FORMER ADMINISTRATOR
DEFENSE TECHNICAL INFORMATION CENTER

# *Preface*

## Purpose and Audience

The purpose of this book is to improve defense-related scientific and technical reports by promoting standardization and by encouraging the creation, distribution, and preservation of scientific and technical information (STI).

The primary audience for this book consists of authors, editors, publishers, and librarians in the defense community. Authors of scientific and technical reports include Department of Defense (DoD) scientists and engineers (civilian and military) and contractors and grantees who support DoD.[a] Editors, as well as illustrators, assist authors and may be responsible for establishing format and style guidelines within a particular organization. Publishers include not only the performing organizations and sponsoring organizations responsible for issuing and funding, respectively, scientific and technical reports, they also include printers, who are responsible for completing and delivering the finished product. Librarians are responsible

---

[a]Unless stated otherwise, references herein to the Government, including the departments and agencies thereof, and the military refer to the U.S. Government and the U.S. Armed Forces, respectively.

for cataloging and preserving scientific and technical reports and for performing literature searches required for research, development, test, and evaluation (RDT&E).

A secondary audience for this book exists among comparable groups outside the immediate defense community but associated with the overall defense establishment, e.g., the Department of Homeland Security and the Department of Energy. These groups may require and seek guidance on certain similar issues, e.g., classified scientific and technical reports. The secondary audience also includes those not part of or affiliated with the Government but who seek guidance on preparing standardized scientific and technical reports within a military framework. This includes State and local government agencies, academia not affiliated with the Government, technical societies, U.S. Allies and friendly foreign governments, and assorted private sector groups and individuals.

## Need for Standardization

It is DoD policy to promote standardization of materiel, facilities, and engineering practices to improve military operational readiness, reduce total ownership costs, and reduce acquisition time. There shall be a single, integrated DSP [Defense Standardization Program] and a uniform series of specifications, standards, and related documents.[1]

Notwithstanding DoD policy, standardization, while ideally sought, is not necessarily the norm among defense-related scientific and technical reports. Numerous variations and inconsistencies exist in format and style among defense-related scientific and technical reports—even within individual DoD and DoD contractor organizations. Also, many defense-related scientific and technical reports do not contain the essential elements of a scientific and technical report or they do not properly arrange these elements.

Standards exist for most types of DoD publications, including scientific and technical reports. ANSI/NISO Z39.18–2005, "Scientific and Technical Reports – Preparation, Presentation, and Preservation,"[2] was published by the National Information Standards Organization (NISO) in 2005 and has been adopted by DoD. However, ANSI/NISO Z39.18–2005 is intentionally brief and general in nature and, therefore, does not totally address the unique, specific requirements of defense-related scientific and technical reports. Moreover, the standard and those that preceded it are not widely known in the defense community and, thus, compliance efforts have been nonexistent in many organizations.

This book addresses and provides guidance on scientific and technical reports in a manner consistent with ANSI/NISO Z39.18-2005. I quote and expand on ANSI/NISO Z39.18-2005, especially those portions of ANSI/NISO Z39.18-2005 requiring clarification on how to prepare defense-related scientific and technical reports, and provide instructions on how to prepare most classified scientific and technical reports. (The contents of this book are unclassified and approved for public release in accordance with Department of Defense Directive [DODD] 5230.9.[3] The guidance on marking classified information is subject to change. Up-to-date instructions may be obtained from the Information Security Oversight Office [ISOO]; ISOO's Web site is www.archives.gov/isoo.)

In addition to and to complement ANSI/NISO Z39.18-2005, I reference and quote other NISO standards cited in ANSI/NISO Z39.18-2005 and related DoD documents. (I was a member of the Standards Committee AH on Scientific and Technical Reports—Elements, Organization, and Design, which developed ANSI/NISO Z39.18-1995, "Scientific and Technical Reports—Elements, Organization, and Design."[4] I also provided comments to the NISO Committee AW regarding ANSI/NISO Z39.18-2005.)

## Creation, Distribution, and Preservation of Scientific and Technical Information

The creation, distribution, and preservation of STI are essential to the defense community in terms of national security and cost effectiveness by reducing if not eliminating the replication of RDT&E. Standardization and STI are intertwined if not synonymous; the former facilitates the usefulness of the latter by allowing it to be presented in a consistent and, thus, efficient and timely manner. Defense-related scientific and technical reports are part of a larger STI network and are one of the principal means of obtaining STI. DoD's Defense Technical Information Center (DTIC) is the central repository for defense-related STI, including scientific and technical reports. Some of the information in this book was obtained from DTIC, whose services are described in Appendix A. DTIC's Web site is www.dtic.mil.

## Approach

I have edited scientific and technical reports since 1988—until 1999 in the Government and since then in industry and now academia—and I have relied heavily on my experience in writing this book.

The NISO standards make voluntary recommendations; they are not dogmatic. Furthermore, while related DoD directives and instructions are usually mandatory, they are not necessarily consistent and oftentimes require interpretation, thereby allowing for a certain amount of latitude on the part of their users. As a result, I have reviewed and assimilated the information in these documents and prescribed what I feel are the "best practices" for preparing a defense-related scientific and technical report within established guidelines.

While most of my experience concerning scientific and technical reports was and continues to be obtained from technically editing scientific and technical reports prepared for the Department of the Navy (DoN), the guidance I provide herein is not necessarily directed to a particular branch of the Armed Forces.

In 1992, the Navy published *STI Handbook: Guidelines for Producing, Using, and Managing Scientific and Technical Information in the Department of the Navy.*[5] The *STI Handbook*, which is referenced in Appendix A of ANSI/NISO Z39.18-2005, interprets and consolidates the plethora of DoN and DoD regulations regarding STI. Certain guidelines have changed since its publication, and Peggy Cathcart, author of the *STI Handbook* and now retired from Government service, asked me if I would be interested in revising and publishing a new handbook. With Peggy's knowledge and concurrence, I opted, instead, to update and incorporate those portions applicable to defense-related scientific and technical reports into this book.

I advocate and present two formats specified in ANSI/NISO Z39.18-2005 for scientific and technical reports: typographical progression[b] and decimal numbering. This book—a hybrid—incorporates features of both formats and may be used as an additional guide on how to prepare defense-related technical reports.

This book is not a specification, but as a former construction specification writer I understand the complementary nature of plans and specifications or, equivalently, visuals and text and the need for both to adequately portray a subject. Thus, I have included many figures in this book, mostly simple, generic drawings showing examples of the various elements of a scientific and technical report. As a result of my DoN experience, some of these examples are derived from or pertain to DoN; however, they should be applicable throughout the defense community. I also include suggested writing and related practices pertaining to tone and style (Appendix B).

This book is an instructional guide and, as such, it is intended to be occasionally referenced without being re-read in its entirety. Thus, certain information in this book has been repeated, at least partially, in other chapters

---

[b]Referred to as "a progression of fonts"[2] in ANSI/NISO Z39.18-2005.

and/or sections of the book to aid the reader; where necessary, related information has been cross-referenced.

In writing this book, I conferred with and received advice from a number of DoD employees; however, neither they nor DoD nor any component of DoD has officially approved, endorsed, or authorized this book. As a result, while certain quotations and referenced directives contain verbs such as "shall" and "must," "should" and "may" are substituted elsewhere to emphasize this book's advisory-only status.

Questions or comments regarding the contents of this book are welcome. My e-mail address is walter.rice@jhuapl.edu.

WALTER W. RICE

# *Acknowledgments*

As I mentioned in the preface, Peggy Cathcart was instrumental in my decision to write this book. Not only did I rely heavily on her book, the *STI Handbook*,[5] for valuable information, but Peggy also favorably evaluated my book proposal and later reviewed and commented on my manuscript.

In addition to Peggy, a number of people provided me with encouragement and support, including my colleagues at the Johns Hopkins University Applied Physics Laboratory (APL). In particular, I would like to thank the members of the APL Subcommittee of the Sabbatical Fellows and Professors Programs for my appointment as a Stuart S. Janney Fellow to pursue the writing and publishing of this book. I would also like to thank the following for their contributions:

- J. Miller Whisnant and Ray Williamson of the Air and Missile Defense Department. Miller reviewed my manuscript and approved it for public release, and Ray provided me with copyright information.
- Susan Fingerman, Heidi Rubinstein, and Christina Pikas of the R. E. Gibson Library and Information Center. All assisted me in my research.
- Ben Roca of the Office of Counsel. Ben assisted me on contractual matters.

I also received assistance outside APL from the following:

- Kurt Molholm wrote the foreword to this book and, as noted there, is a former DTIC Administrator and Chair of the NISO Committee AW for ANSI/NISO Z39.18-2005.[2] Kurt also kept me abreast of the status of the ANSI/NISO Z39.18 revision.
- Pat Harris is a former executive director of NISO. I worked with Pat during the 1995 revision to ANSI/NISO Z39.18, and she encouraged me to comment on the 2005 revision. Pat granted me permission to quote and excerpt all versions of ANSI/NISO Z39.18 and other NISO standards referenced in ANSI/NISO Z39.18.
- Sharon Serzan and Brian Hermes are friends and colleagues who reviewed my manuscript and provided comments. My wife, Myra Rice, also reviewed my manuscript and provided comments.
- Betty Fomin is an illustrator and software instructor and owner of Grantec, Inc. Betty helped design my book cover and assisted me with formatting issues.
- Bill Wells is a retired lawyer. Bill, my uncle, also assisted me on contractual matters.

# Chapter *1*

# *Background*

## 1.1    Definition of a Technical Report

In the scientific and technical community, the term "report" is applied to a number of documents that, oftentimes, have common elements, such as an abstract, and share a collective purpose—to inform a specific audience about a particular subject. They range from the formal to the informal. However, a traditional scientific and technical report or, simply, technical report, as it is commonly known, is unique among reports. ("Scientific and technical report" and "technical report" are used interchangeably among the various scientific and technical disciplines and in the related standards; the terms are essentially synonymous. For brevity, "technical report" will be used hereinafter.)

By definition, technical reports "convey the results of basic or applied research and support decisions based on those results. A [technical] report includes the ancillary information necessary for interpreting, applying, and replicating the results or techniques of an investigation. The primary purposes of such a [technical] report are to disseminate the results of scientific and technical research and to recommend action."[4] (As explained later in this chapter, the term "technical report" has become more inclusive and less restrictive and includes documents that do not necessarily fit this definition.)

A technical report is similar to a technical or scientific paper. Both are an integral and essential part of research, and, by definition, both should provide enough information so that an experienced and knowledgeable investigator in the same field could repeat the described procedure and obtain the same results.

The differences between a technical report and a scientific paper are few but noteworthy. A technical report often has a limited distribution within an organization and, therefore, a limited readership. A scientific paper is usually published in a journal, which, for the most part, is publicly accessible, and it may be presented orally at a conference. (Notwithstanding, there are some journals that publish classified scientific papers.) Technical reports tend to be more extensive than scientific papers because of page restrictions in journals. Occasionally, some technical reports are submitted to journals for publication as scientific papers.

While the accompanying information contained in the front and back matter of technical reports distinguishes them from scientific papers, the core elements of technical reports and scientific papers are closely related, i.e., they contain an introduction, a section on materials and methods used, a section discussing the results obtained, and a section presenting the conclusions reached. However, a technical report reflects an organizational effort, as opposed to an individual or personal effort. As such, it should additionally contain a summary (in addition to an abstract) or, as it is more commonly known, an executive summary, which is intended to read by management within the particular organization sponsoring the work, and a recommendations section, which prescribes a future course of action for decisionmakers within the same organization.

Technical manuals, which are commonplace in the defense community, "are publications that contain instructions for the installation, operation, maintenance, training, and support of weapon systems, weapon system components and support equipment."[6] Technical manuals should not be confused with or presented as technical reports, which serve a different purpose. Occasionally, portions of a technical manual may be incorporated into or appended to a technical report. This usually occurs when, as a result of research findings, a technical manual requires modification.

Warnings, cautions, and notes frequently appear in technical manuals to highlight certain information; their use in technical reports is rare and should only occur when they are part of an incorporated or appended technical manual. They should not be used to highlight other portions of a technical report. Further guidance on warnings, cautions, and notes is provided in MIL-STD-38784.[6] Figure 1-1, derived from MIL-STD-38784, shows a warning, caution, and note and provides a description of each.

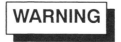

Highlights an essential operating or maintenance procedure, practice, condition, statement, etc., which, if not strictly observed, could result in injury to, or death of, personnel or long term health hazards.

Highlights an essential operating or maintenance procedure, practice, condition, statement, etc., which, if not strictly observed, could result in damage to, or destruction of, equipment or loss of mission effectiveness.

**NOTE**

Highlights an essential operating or maintenance procedure, condition, or statement.

**Figure 1-1.   Warning, Caution, and Note**

Memorandums are often incorrectly substituted for technical reports. By definition, a memorandum is an internal document. A memorandum may appear similar to a technical report and eventually evolve into a technical report, but because of its transitory nature it does not, in itself, merit the significance or retain the permanence of a technical report.

Other technical documents "such as brochures, proposals, reliability plans, safety plans, viewgraph compilations, conference proceedings, computer programs, engineering change proposals, and specifications"[7] also serve a different purpose than technical reports and should not be confused with or presented as such.

Because of category limitations on the various types of technical documents, some are often loosely considered technical reports and some, in fact, have been included in databases such as the Defense Technical Information Center's (DTIC's) technical report database. Regardless, a distinction should be made between technical reports and other technical documents.

The definition of a technical report has, in itself, not always been standard and has changed over time, as have the various standards addressing technical reports.

## 1.2     Standards

## 1.2.1     History

The Department of Defense (DoD) began an initiative years ago to replace, where possible, military standards with industry standards. Originally, MIL-STD-847, "Format Requirements for Scientific and Technical Reports Prepared by or for the Department of Defense,"[8] established guidelines on how to prepare a technical report for DoD. MIL-STD-847 is a relatively brief document that is less elaborative than its successor industry standards and now clearly dated; it does, however, provide some guidance regarding classified technical reports.

MIL-STD-847 defines a technical report as a document that "encompasses the evaluated relevant facts on a study or phase of a study of a particular art, science, profession, or trade, and stands as a permanent official record in a formal document."[8] Interestingly, this definition of a technical report may more accurately describe what a technical report is currently considered, as discussed later in this chapter. The absence of a phrase regarding the need for "ancillary information necessary for interpreting, applying, and replicating the results or techniques of an investigation"[4] is noteworthy and indicates the evolutionary nature of the definition of a technical report that would eventually be changed when an industry standard superseded the military standard.

ANSI Z39.18-1987, "Scientific and Technical Reports - Organization, Preparation, and Production,"[9] which superseded MIL-STD-847, was sponsored by the National Information Standards Organization (NISO), approved by the American National Standards Institute (ANSI), and adopted by DoD. ANSI Z39.18-1987 defines a technical report as "a publication designed to convey the results of basic or applied research."[9] ANSI Z39.18-1987 introduces the statement that a technical report includes "the ancillary information necessary for the interpretation, application, and replication of the results or techniques of an investigation."[9] However, it does not contain the phrase "and support decisions based on those results."[4] This phrase, which highlights the role of management in technical reporting, was added in ANSI/NISO Z39.18-1995.

ANSI/NISO Z39.18-1995, "Scientific and Technical Reports—Elements, Organization, and Design,"[4] was developed by NISO, approved by ANSI, and adopted by DoD. ANSI/NISO Z39.18-1995 closely resembles its predecessor, ANSI Z39.18-1987, in layout and appearance. ANSI/NISO Z39.18-1995 "establishes guidelines for the elements, organization, and design of scientific and technical reports."[4] The guidelines in ANSI/NISO Z39.18-1995 pertain exclusively to print (paper and ink) technical reports, and, as such, ANSI/NISO Z39.18-1995 soon became limited in its application with the increasing number of technical reports being prepared and presented electronically.

The standard on technical reports entered the electronic or digital age with the issuance of ANSI/NISO Z39.18-2005, "Scientific and Technical Reports - Preparation, Presentation, and Preservation."[2]

### 1.2.2    ANSI/NISO  Z39.18-2005,  "Scientific  and  Technical Reports - Preparation, Presentation, and Preservation"

As stated in the preface, ANSI/NISO Z39.18-2005 is the current DoD standard for technical reports, and, like its predecessor, it was developed by NISO and approved by ANSI. ANSI/NISO Z39.18-2005 "outlines the elements, organization, and design of scientific and technical reports, including guidance for uniform presentation of front and back matter, text, and visual and tabular material in print and digital formats, as well as recommendations for multimedia [technical] reports."[2]

While continuing to provide guidance that closely resembles ANSI/NISO Z39.18-1995 on how to present print technical reports, ANSI/NISO Z39.18-2005 is an otherwise significant departure from its predecessor. ANSI/NISO Z39.18-2005 discusses the evolution of the organization of a technical report from "a content-based to a user-based organizational pattern."[2] As part of this discussion, the standard introduces "metadata" to the standard. Metadata "refer to information about information or, equivalently, data about data."[2] A thorough description of metadata is provided by NISO in "Understanding Metadata."[10]

To further define the term, ANSI/NISO Z39.18-2005 draws a distinction between content and metadata: "The body of the [technical] report, with its discussion of methods, results, and conclusions, is content. Any information that helps the user find, assemble, and properly attribute the [technical] report is metadata."[2]

ANSI/NISO Z39.18-2005 emphasizes the significance of metadata and describes their three classes: descriptive, structural, and administrative.

Metadata are a significant matter . . . because of the large amount and diversity of data represented. The quantity and diversity of [technical] report content and format presented information management challenges in an era when [technical] reports were published exclusively on paper; in the digital age these challenges have multiplied considerably. A scientific [and] technical report that does not take metadata into account has no readily-found identify and will not be used. To avoid this problem, compilers of [technical] reports must provide metadata in three broad classes: descriptive, structural, and administrative.

Descriptive metadata . . . convey information that helps the user find a [technical] report and distinguish it from other similar ones. . . . Such metadata include the title and creator (author), as well as any keywords or subject references.

Structural metadata explain the relationship between parts of multipart objects and enhance internal navigation. Such metadata include a table of contents or list of figures and tables.

Administrative metadata support maintaining and archiving [technical] reports and ensure their long-term availability. . . . Such metadata include type and version of software used in preparing the [technical] report and rights-management requirements.[2]

Descriptive and structural metadata are integral parts of any good technical report—print or digital. Administrative data relate, primarily, to digital technical reports.

Acknowledging if not acquiescing to the evolving nature of what constitutes a "technical report," ANSI/NISO Z39.18-2005 does not attempt to define a technical report. Furthermore, it also states that "there is no single format [regarding technical reports] that authors should use."[2] Notwithstanding, ANSI/NISO Z39.18-2005 basically repeats the conventional format for technical reports presented by ANSI/NISO Z39.18-1995; however, it also prescribes optional, shortened formats.

ANSI/NISO Z39.18-2005 discusses "print-specific"[2] (paper and ink) technical reports versus "non-print-specific"[2] (totally digital) technical reports, the implication seemingly being that print technical reports may be "born digital"[2] or created electronically but, ultimately, are meant to be printed and, thus, read on paper. (Today, technical reports are rarely, if ever, hand-written or, for that matter, produced on a typewriter.) Non-print technical reports are "born digital"[2] and remain digital and are intended to be read or viewed almost exclusively on screen and, in some instances, heard. Non-print technical reports may comprise multimedia and employ features unavailable to print technical reports, such as sound and animation. They, ultimately, present a distinct contrast to the common features of print technical reports.

The guidance provided herein prescribes two conventional formats applicable to any print defense-related technical report, regardless if it completely satisfies the traditional definition presented earlier in this chapter. While the logic employed in presenting print technical reports may, in certain instances, be carried over to non-print technical reports, the latter, nevertheless, remain the exception, not the rule, in the defense community. Furthermore, to attempt to address all facets of non-print defense-related technical reports would be challenging if not overwhelming and is, thus, beyond the scope of this effort.

### 1.2.3    Related NISO Standards

NISO has developed two other standards related to technical reports: ANSI/NISO Z39.23-1997[11] and ANSI/NISO Z39.14-1997.[12]

### 1.2.3.1    ANSI/NISO    Z39.23-1997,    "Standard    Technical Report Number Format and Creation"[c]

ANSI/NISO Z39.23-1997 provides formatting guidance on how to create a standard technical report number. The same subject is addressed herein, and an example of a typical DoD organizational technical report number is provided. ANSI/NISO Z39.23-1997 is referenced in NISO Z39.18-2005 and, therefore, is part of that standard.

### 1.2.3.2    ANSI/NISO Z39.14-1997, "Guidelines for Abstracts"[c]

ANSI/NISO Z39.14-1997 provides guidance on how to write an abstract, which is a required element of a technical report. Abstracts are addressed herein, and visual examples of typical abstracts appearing in defense-related technical reports are provided. ANSI/NISO Z39.23-1997 is referenced in NISO Z39.18-2005 and, therefore, is part of that standard.

ANSI/NISO Z39.18-2005, ANSI/NISO Z39.23-1997, and ANSI/NISO Z39.14-1997 may be downloaded and printed free of charge at the following Web site: www.niso.org or purchased through NISO.

---

[c]Revised in 2002.

## 1.3    Department of Defense's Scientific and Technical Information Program

DoD has established and maintains a scientific and technical information program (STIP) "to document the results and outcome of DoD-sponsored and/or performed research and engineering (R&E) and studies efforts and provide access to those efforts in an effective manner consistent with the DoD mission."[13] As a source of documentation, technical reports are an essential if not the main component of DoD's STIP. DoD attaches special significance to its STIP:

> The Department of Defense shall aggressively pursue a coordinated and comprehensive STIP, thereby providing maximum contribution to the advancement of science and technology. The STIP shall permit timely, effective, and efficient conduct and management of DoD research and engineering (R&E) and studies programs, and eliminate unnecessary duplication of effort and resources by encouraging and expediting the interchange and use of STI [scientific and technical information]. Interchange and use of DoD STI is intended to include the DoD Components, their contractors, other Federal Agencies, their contractors, and the national and international R&E community.[14]

Scientists and engineers in the defense community should be aware of their role in contributing to DoD's STIP. However, safeguards should be maintained. Sections 1.4 to 1.7 provide an overview of the responsibilities of those involved in the process, as well as a discussion of the various means of controlling STI.

## 1.4    Publishing Requirements[d]

Research, development, test, and evaluation (RDT&E) is the general, inclusive category under which the defense community produces STI. Publication of RDT&E in the form of a technical report is a fundamental precept in the scientific and engineering profession. As such, it attests to the professionalism of one's organization and increases the stature of its scientists and engineers.

Technical reports are often the only RDT&E products seen by high level military personnel and congressional groups. RDT&E not reported is basically irrelevant. Technical reports assist congressional committees and DoD

---

[d]Most of the information in Sections 1.4 to 1.6 was obtained from the *STI Handbook*.[5] The information has been updated, expanded, and adapted to apply to technical reports, in particular, as opposed to STI, in general.

organizations plan how to allocate funding for RDT&E and determine what activities should receive funding. Conversely, lack of published RDT&E can decrease available funding.

Duplication of RDT&E causes loss of research funds. DoD considers this wasteful and possibly a matter to be investigated by an inspector general. An audit trail of how public funds have been spent is provided within the technical report. Because a technical report has been reported to DTIC, a permanent record of RDT&E has been established and DoD's expenditure of public funds has been maintained.

### 1.4.1    Author Responsibilities

Scientists and engineers (authors) in the defense community should comply with the following publishing requirements:

- DoD scientists and engineers, as well as those scientists and engineers under contract to DoD (industry and academia), should publish the results of RDT&E in the form of a technical report and submit the technical report to DTIC for inclusion in its technical report database.
- Positive and negative results should be published. Specifically, defense community scientists and engineers should ensure that all significant scientific or technological observations, findings, recommendations, and results derived from DoD endeavors, including those generated under contracts or grants pertinent to the DoD mission, that contribute to the DoD or national scientific or technological base are recorded.
- Technical reports should be completed in a timely manner, usually within 6 months of the conclusion of the RDT&E or within 6 months of a significant scientific or technological observation. Draft technical reports should not languish so that their relevance is diminished if not obsolete.
- Scientists and engineers should budget for the costs of editing and publishing their technical reports. Technical editors and illustrators should, if possible, assist authors in the preparation of technical reports. The involvement of technical editors and illustrators is usually dependent upon authors. While authors are inclined to seek the help of illustrators, they are seemingly reluctant to approach technical editors for assistance. The primary job of a technical editor is to evaluate not what is being presented in a technical report, which is the responsibility of a technical reviewer (Section 1.5.1), but how it is being presented. Technical editors usually become involved after a draft has been completed; however, they should begin advising au-

thors at, preferably, an earlier stage in the technical report process to alleviate potential problem areas.

## 1.4.2    Non-Author Responsibilities

In addition to the authors' responsibilities, others have certain responsibilities with regard to publishing RDT&E:

- Supervisors should ensure that engineers and scientists under their supervision and those of their DoD contractors publish their RDT&E in the form of a technical report.
- Project and program managers should ensure that the results of their projects and programs, respectively, are published in the form of a technical report. They should also budget for and fund the publication of technical reports and coordinate their respective project's and program's requirements.
- A contracting officer's technical representative (COTR) should ensure the results of contract RDT&E are published in the form of a technical report. Acquisition regulations require that DoD contractors provide technical reports, consistent with the objectives of the RDT&E involved, as a permanent record of the RDT&E accomplished under the contract. COTRs should also include applicable data item descriptions (DIDs) for a technical report in the contract and ensure that DoD contractors are aware of their responsibilities.

## 1.5    Review of Technical Reports

All technical reports should be reviewed by someone other than the author before publication; however, technical reports intended for public release are required to be reviewed to ensure that they are suitable for public release and, thus, have the correct distribution statement. The review process usually includes the public affairs office of the performing organization. The individual responsible for releasing a technical report to the public should be specifically authorized by the appropriate officials to perform this function.

The purpose of the total review process is to do the following:

- Check the technical report for technical adequacy;
- Verify the classification level of the technical report and, if the technical report is classified, to ensure all portions are properly marked and all applicable notices are included;
- Select the appropriate distribution statement;
- Determine if export-controlled information has been included in the technical report; and
- Check for patentable information.

## 1.5.1     Technical Review

The purpose of a technical review is to improve the value of the technical report by evaluating the scientific and technical merit of the work. The role of a technical reviewer is to ascertain the technical adequacy of a technical report. As such, the basic concerns of a technical reviewer are to ensure accurate reporting of the material, reasonable interpretation of the data, and logical inferences.

Supervisors usually perform a technical review; however, an independent review of the technical report may be performed by another engineer or scientist who is a subject matter expert. The technical content is evaluated, and ways are suggested to improve communication with the intended audience. While comments regarding non-technical matters, e.g., grammar and punctuation, are usually welcome, technical reviews should not concentrate on these aspects of a technical report; they are best left to a technical editor (Section 1.4.1).

A technical reviewer should remain aware of the status of the technology being reported and should, if possible, do the following:

- Become a subject matter expert on the status of the technology in the defense community, including industry and academia, and the international arena;
- Ask the internal security office for assistance in interpreting security guidelines;
- Use DTIC and the organization's library or technical information center (TIC) for resources and should, ideally, be a regular contributor to both;
- Stay aware of independent research and development (IR&D) in academia and industry;
- When applicable, use the local intelligence office to obtain STI on military equipment belonging to U.S. Allies and adversaries;
- Participate in technical professional societies; and
- Use the Militarily Critical Technologies List and provide input to DoD to change the list as technology changes.

## 1.5.2     Security Review

The author usually makes the initial determination concerning the classification of the technical report. If the author is in doubt about its classification, the organization's security office should be consulted. In deciding if the information in the technical report is classified, the following are areas that might require classification:

- Performance and capabilities

- Specifications
- Vulnerabilities
- Procurement and production plans and schedules
- Operations.

If the technical report is classified, all portions should be properly marked, the correct classification source should be cited, and the declassification or review information should be correct. If multiple sources are used to classify the technical report, a list of those sources should be included with the technical report. Classified technical reports are discussed in greater detail in Section 1.7.

## 1.5.3    Distribution Statement Review

The author should make the initial determination concerning what organizations can have access to the technical report. This decision then forms the basis for selecting the correct distribution statement. In making the decision, the author tentatively determines if the technical report can be released to the public without any restrictions; if so, the technical report is in the public domain, where it can be appropriated by anyone, including foreign nationals or representatives of foreign industries or governments.

DoD has the following seven broad RDT&E categories:

- Basic research
- Applied research
- Advanced technology development
- Advanced component development and prototypes
- System development and demonstration
- RDT&E management support
- Operational systems development.[15]

Two of the RDT&E categories affect whether a technical report may be unlimited or limited in its distribution: basic research and applied research.

Basic research is systematic study directed toward greater knowledge or understanding of the fundamental aspects of phenomena and of observable facts without specific applications towards processes or products in mind. It includes a scientific study and experimentation directed toward increasing fundamental knowledge and understanding in those fields of the physical, engineering, environmental, and life sciences related to long-term national security needs. It is farsighted high payoff research that provides the basis for technological progress. Basic research may lead to: (a) subsequent applied research and advanced technology developments in Defense-related technologies, and (b) new and improved military financial capabilities in areas such as communications, detection, tracking, surveillance, propulsion, mobility, guidance and control, navigation, energy conversion, materials and structures, and personnel support.

Applied research is systematic study to understand the means to meet a recognized and specific need. It is a systematic expansion and application of knowledge to develop useful materials, devices, and systems or methods. It may be oriented, ultimately, to the design, development, and improvement of prototypes and new processes to meet general mission area requirements. Applied research may translate promising basic research into solutions for broadly defined military needs, short of system development. This type of effort may vary from systematic mission-directed research beyond that in [basic research] to sophisticated breadboard hardware, study, programming and planning efforts that establish the initial feasibility and practicality of proposed solutions to technological challenges. It includes studies, investigations, and non-system specific technology efforts. The dominant characteristic is that applied research is directed toward general military needs with a view toward developing and evaluating the feasibility and practicality of proposed solutions and determining their parameters.[15]

To help determine if work performed under basic research or applied research can be released to the public, basic research may usually be released to the public until the "state of emergence"[5] is evident, i.e., the transition has been made from basic research to applied research with specific military applications. After that has occurred, the distribution of the technical report should be limited. (The distinction between basic research and applied research is a sensitive if not controversial issue in the defense community; a thorough review process should be conducted before releasing a technical report to the public.)

If the technical report cannot be released to the public, the author should determine the extent of its availability without additional approval of the controlling organization. The controlling organization is usually the sponsoring organization or the preparing organization. Distribution statements allow the author to limit access of the technical report to any of the following groups:

- U.S. Government agencies
- U.S. Government agencies and their contractors
- DoD and U.S. DoD contractors
- DoD components
- U.S. Government agencies and private individuals or enterprises eligible to obtain export-controlled technical data.

In addition, the author can completely limit access to the technical report and require that all requests for the technical report be approved by the controlling organization.

After the author has decided on the level of access, the author should select a reason for the limitation. Reasons for limiting access to a technical report are as follows:

- Foreign government information
- Proprietary information
- Critical technology
- Test and evaluation
- Contractor performance evaluation
- Premature dissemination
- Administrative/operational use
- Software documentation
- Specific authority
- Direct military support.

An author's decision to select a certain distribution statement should be reviewed and approved by the author's supervisors and any other internal office that makes the final determination regarding distribution statements. Technical reports approved for public release should be reviewed and approved by the appropriate DoD component. Distribution statements are discussed in greater detail in Chapter 3, Section 3.12.6.

### 1.5.4    Export Control Review

An author should make a decision regarding export controls when determining a distribution statement. When making the decision, the author should consider DoD's Militarily Critical Technologies List, the Department of State's U.S. Munitions List, the Department of Commerce's Commerce Control List, and the author's knowledge concerning the state of the art of the applicable technology. Export controls are discussed in greater detail in Chapter 3, Section 3.12.7.

### 1.5.5    Patent and Technology Transfer Review

All defense-related technical reports should be reviewed for inclusion of STI that can be patented and, if there is commercial interest, transferable as a technology to business and industry. Financial incentives are often put in place by various organizations for authors who produce such STI. The legal counsel of the author's organization should perform the patent review. (Most large organizations retain patent counsels.) Some large organizations maintain a technology transfer office, which should be able to review a technical report for applicable STI.

## 1.6    Distribution of Technical Reports

### 1.6.1    Basic Guidelines

When distributing a technical report, a recipient other than DTIC and, perhaps, an organization's library or TIC should, at a minimum, receive one paper copy. (DTIC's distribution requirements are further discussed in Section 1.6.2.2.1 and Appendix A, Section A.2.1.2; organizational distribution is discussed in Section 1.6.2.1.) A digital or electronic copy is provided if an organization's library or TIC specifically requests it to facilitate reproduction and preservation or a portion of the technical report is solely electronic in nature, e.g., a multimedia appendix or a software program; otherwise, an electronic copy of the technical report is optional. Occasionally, a technical report comprises a combination of the two.

With the exception of DTIC and an organization's library or TIC, caution should be exercised when distributing electronic copies of final technical reports. Paper copies are more difficult and, therefore, less likely to be altered and reproduced without the author's knowledge and approval; thus, the author retains greater control over the technical report's integrity and distribution. Electronic copies of final technical reports should be saved in an unalterable or read-only status before they are distributed.

Electronic copies of technical reports may be distributed in a number of formats; however, a Portable Document Format (PDF), especially if the software used to create the technical report is uncommon, is generally the preferred format. The International Organization for Standardization (ISO) published a standard in 2005 for archiving PDF files, called PDF/A-1,[16] which should facilitate the long-term preservation of electronic technical reports. The standard may be obtained from ISO. ISO's Web site is www.iso.ch/iso/en/ISOOnline.

The preparing organization is responsible for distributing the technical report to certain individuals. Only personnel with a need to know and an appropriate security clearance should receive a copy of classified and unclassified, limited distribution technical reports. All technical reports should be properly reviewed to assign the correct distribution statement, determine the classification level, and verify their technical adequacy.

### 1.6.2    Primary Distribution

Primary distribution is the initial distribution (internal and external) of technical reports, i.e., those copies of a final technical report distributed immediately after printing or after an electronic copy of the technical report has been made.

## 1.6.2.1    Internal Distribution

Internal distribution refers to distribution to personnel within an author's organization. In addition to copies for the author and any co-authors, the internal distribution list should include the following: the author's chain of command up through the head of the organization, colleagues of the author who might have an interest in the technical report, the organization's library or TIC (reference and archival use), and the organization's legal counsel (patent counsel). Stock paper copies should be created for anticipated secondary distribution; however, large quantities should be unnecessary because of the ease of electronic reproduction.

## 1.6.2.2    External Distribution

### 1.6.2.2.1    General Guidelines

External distribution refers to distribution to individuals and organization's outside one's own organization. This includes DTIC, sponsors, other DoD organizations, and DoD contractors, including academia. The following guidelines for external distribution should be used:

a.  The Government-Industry Data Exchange Program (GIDEP) should be provided a copy. GIDEP is managed by DoN; however, it encompasses the entire defense community and is "a cooperative activity between government and industry participants seeking to reduce or eliminate expenditures of resources by sharing technical information essential during research, design, development, production and operational phases of the life cycle of systems, facilities and equipment."[17] Classified technical reports and technical reports that contain proprietary information should not be sent to GIDEP. GIDEP's Web site is www.gidep.org.

b.  If the technical report is classified, the security clearances of any contractors on the primary distribution list should be verified to ensure they can receive classified information up to the overall classification level of the technical report and they have a need to know the information revealed in the technical report. Verification is done through DoD's Defense Security Service (DSS). DSS's Web site is www.dss.mil.

c.  If the technical report contains export-controlled information, the eligibility of any contractors on the primary distribution list to receive specific export-controlled information should be verified. Verification is done through the Defense Logistics Agency, specifically the

Defense Logistics Information Service's United States/Canada Joint Certification Program (JCP). JCP's Web site is www.dlis.dla.mil/ JCP.

d. Transmittal letters should be prepared if the author's organization requires transmittal letters for technical reports sent to other organizations. Classified technical reports should follow the appropriate instructions regarding mailing classified information.

### 1.6.2.2.2    Secondary Distribution

Secondary distribution refers to requests received after primary distribution is completed. Most secondary distribution requests come either in response to DTIC's announcement and inclusion of the technical report in its technical report database or through the use of the technical report as a reference. Although a performing organization may receive a request for secondary distribution of a technical report, it is the responsibility of DTIC to perform this task. Requesters, except those within the author's organization, should be referred to DTIC.

One of the principal functions of DTIC is to acquire, store, announce, retrieve, and provide secondary distribution of DoD STI. DTIC services are available to all components of DoD, DoD contractors, other Government organizations and their contractors, grantees, and potential contractors. Before DTIC can release a limited distribution technical report to someone not authorized by the assigned distribution statement to receive the technical report, DTIC must have the approval of the controlling (sponsoring) organization identified in the distribution statement. Approval or disapproval is obtained through the use of DTIC Form 55. If the technical report is classified or subject to export control regulations, DTIC verifies that the requester has the appropriate clearances, i.e., security and export control registration, before forwarding DTIC Form 55 to the controlling organization. Figures 1-2 and 1-3 show DTIC Form 55. Figure 1-2 shows the form itself; Figure 1-3 shows the instructions for completing it. DTIC Form 55 is available online at www.dtic.mil/dtic/forms/user_form55.html.

### 1.6.2.2.3    Foreign Distribution

Distribution of technical reports to foreign nationals is strictly controlled. The performing organization's security office should be consulted before releasing a technical report to a foreign national.

Technical reports that have been cleared for public release may be released to foreign nationals. Foreign nationals may be included on the primary distribution list in technical reports that have been cleared for public

| DEFENSE TECHNICAL INFORMATION CENTER<br>REQUEST FOR RELEASE OF LIMITED DOCUMENT | DTIC CONTROL NO. | USER ROUTING |
|---|---|---|

SECTION I - REQUESTING INFORMATION

| 1. REQUESTING ORGANIZATION AND ADDRESS: | 2. DTIC USER CODE NO. | 3. DATE OF REQUEST |
|---|---|---|
| | 4. TYPE COPY AND QUANTITY<br>☐ Paper Copy ___ Copy(s)  ☐ Microfiche ___ Copy(s) | |
| | 5. CONTRACT NUMBER | 6. CONTRACT SECURITY LEVEL |
| 7. GOVERNMENT SPONSOR AND ADDRESS (Contractors and Grantees Only) | 8. METHOD OF PAYMENT (X ONE) Acct No. _____<br>☐ VISA ☐ MC ☐ AMEX  Expires: _____<br>☐ Charge to my NTIS Deposit Account No: _____ | |
| 9. CONTRACT MONITOR AND TELEPHONE NUMBER (Contractors and Grantees Only) | 10. NAME, TITLE, TELEPHONE OF REQUESTING OFFICIAL | |
| | EMAIL | FAX NO. |

SECTION II - BIBLIOGRAPHIC INFORMATION

11. AD NUMBER (If known)

12. TITLE, REPORT NUMBER, AUTHOR(S)

SECTION III - REQUESTER JUSTIFICATION

13. REQUESTER JUSTIFICATION (Explain need in detail)

SECTION IV - RELEASING AGENCY

| 1. RELEASING AGENCY ADDRESS (If known) | 2. RELEASING AGENCY DECISION (If the report was developed under the SBIR Program, refer to instruction B.8)<br>☐ APPROVED FOR RELEASE TO THE ABOVE REQUESTER<br>☐ DISAPPROVED. REASON FOR DISAPPROVAL<br>☐ APPROVED FOR PUBLIC RELEASE<br>☐ DISTRIBUTION AUTHORIZED TO U.S. GOVT AGENCIES AND THEIR CONTRACTORS<br>☐ DISTRIBUTION AUTHORIZED TO U.S. GOVT AGENCIES ONLY<br>☐ DISTRIBUTION AUTHORIZED TO DOD ONLY<br>☐ DISTRIBUTION AUTHORIZED TO DOD/ THEIR CONTRACTORS | | |
|---|---|---|---|
| FAX NUMBER | | | |
| EMAIL ADDRESS | | | |
| 3. NAME/TITLE OF RELEASING OFFICIAL | TEL NO. | 5. SIGNATURE | 6. DATE |

DTIC FORM 55 FEB 2004 (EG)                    PREVIOUS EDITIONS ARE OBSOLETE

Figure 1-2.   DTIC Form 55 (Page 1 of 2)

---

**DTIC - FORM 55 INSTRUCTIONS**

**A. DTIC REQUESTER (Complete Sections I, II, and III)**

1.  Enter your routing information in the User Routing block, if desired, for your internal control purposes.

2.  Contractors and Grantees must identify in Section I their government sponsor's name and telephone number, for need-to-know purposes. Please also provide FAX number and email address.

3.  Separate Form 55's must be completed for each request, unless the Releasing Agency is the same for all AD numbers requested.

4.  Explain in detail your requirement for the document. Include appropriate contract information and explain need-to-know in Section III.

5.  Method of payment is required. Retain a copy for your records, mail or fax to:

    DEFENSE TECHNICAL INFORMATION CENTER
    ATTN: DTIC-BC (Registration)
    8725 JOHN J. KINGMAN ROAD, SUITE 0944
    FORT BELVOIR, VA 22060-6218

    Commercial:   703-767-8271          DSN:     427-8271
    FAX:          703-767-9459          DSN:     427-9459

DTIC will not accept any form of prepayment with this request.
(Service charge will be made only for documents approved for release.)

**B. RELEASING AGENCY (Complete Section IV)**

1.  Contractor's Government Sponsor's address, name and telephone number is included in Section I (Blocks 7 & 9) for your use.

2.  Indicate in Section IV, (Block 2) approval or disapproval. Also check the appropriate block, if the distribution statement should be changed.

3.  It is important to complete blocks 3-6. DTIC cannot process Form 55's without a signature.

4.  Please provide your FAX number and email address.

5.  Retain a copy for your records, mail or fax a copy to:

    DEFENSE TECHNICAL INFORMATION CENTER
    ATTN: DTIC-BC (Registration)
    8725 JOHN J. KINGMAN ROAD, SUITE 0944
    FORT BELVOIR, VA 22060-6218

    Commercial:   703-767-8271          DSN:     427-8271
    FAX:          703-767-9459          DSN:     427-9459

6.  Any documents needed for review can be provided free of charge. DTIC policy requires a memo for Code 5 documents (Further Dissemination Only) stating that the document is needed for review. Classified documents require a DTIC User Code before they can be ordered.

7.  As directed by ODDR&E (AT/L), Releasing Agencies should complete the form and return it to DTIC within 15 days.

8.  **WARNING;** If the requested information is proprietary data developed under a **SBIR contract**, it *cannot* be released outside the U.S. Government for a period of FIVE years, after acceptance of the last contract deliverable item, without the written permission of the contractor (DFAS 252-227-7018).

---

**Figure 1-3.   DTIC Form 55 (Page 2 of 2)**

release after approval by the performing organization's security office; secondary distribution requests should be sent to the performing organization's security office.

The release of unclassified, limited distribution and classified technical reports to foreign nationals should be cleared by DoD. Requests to include a foreign national on the initial distribution list on unclassified, limited distribution and classified technical reports are evaluated by DoD on a case-by-case basis; primary and secondary distribution requests should be forwarded to the performing organization's security office, which will coordinate the request with DoD.

## 1.7    Classified Technical Reports

### 1.7.1    Description

Classified technical reports contain classified information, which is "information that has been determined pursuant to this order [Executive Order 13292] or any predecessor order to require protection against unauthorized disclosure and is marked to indicate its classified status when in documentary form."[18] While some foreign countries have additional levels of classification, there are only three levels of classification for U.S. defense-related technical reports: Top Secret, Secret, and Confidential.

   a.   Top Secret "shall be applied to information, the unauthorized disclosure of which reasonably could be expected to cause exceptionally grave damage to the national security that the original classification authority is able to identify or describe."[18] Examples include information whose unauthorized release could result in armed hostilities against the United States or its Allies, a disruption of foreign relations vitally affecting the national security, the compromise of vital national defense plans, the disclosure of complex cryptographic and communications intelligence systems, the disclosure of sensitive intelligence operations, and the disclosure of significant scientific or technological developments vital to national security.[19]

   b.   Secret "shall be applied to information, the unauthorized disclosure of which reasonably could be expected to cause serious damage to the national security that the original classification authority is able to identify or describe."[18] Examples include information whose unauthorized release could result in the disruption of foreign relations significantly affecting the national security, the significant impairment of a program or policy directly related to the national security, the disclosure of significant military plans or intelligence operations, and the disclosure of scientific or technological developments relating to national security.[19]

c. Confidential "shall be applied to information, the unauthorized disclosure of which reasonably could be expected to cause damage to the national security that the original classification authority is able to identify or describe."[18] Examples include information whose unauthorized release could result in disclosure of ground, air, and naval forces (e.g., force levels and force dispositions) or disclosure of performance characteristics such as design, test, and production data of U.S. munitions and weapon systems.[19]

Most defense-related technical reports are unclassified. A technical report should not be considered for classification unless it concerns one or more of the following areas listed in Executive Order 13292[18]:

a. Military plans, weapons systems, or operations;

b. Foreign government information;

c. Intelligence activities (including special activities), intelligence sources or methods, or cryptology;

d. Foreign relations or foreign activities of the United States, including confidential sources;

e. Scientific, technological, or economic matters relating to the national security, which includes defense against transnational terrorism;

f. U.S. Government programs for safeguarding nuclear materials or facilities;

g. Vulnerabilities or capabilities of systems, installations, infrastructures, projects, plans, or protection services relating to the national security, which includes defense against transnational terrorism; or

h. Weapons of mass destruction.

## 1.7.2    Author Responsibilities

An author of a classified technical report is responsible for the following:

• Correctly marking the technical report, including electronic copies, so that no doubt exists about its overall level of classification;

• Correctly marking pages, headings, paragraphs, lists, figures and tables, figure and table titles, and footnotes in the technical report;

• Determining how long the technical report should remain classified based on the classification instructions pertinent to the subject matter;

• Incorporating any additional measures necessary to protect the technical report, e.g., identifying export-controlled information; and

• Identifying the originator of the technical report (usually the preparing organization).

### 1.7.3    Original and Derivative Classification

Original classification involves determining that the material requires protection from unauthorized disclosure and the level of protection required. These original classification determinations are usually issued as program security classification guides. Subsequently, any time this material is used in any form, it is derivatively classified based on the original classification determination.

Derivative classification is done by anyone who incorporates, paraphrases, restates, or generates in new form material that is already classified. Derivative classification is most commonly done by marking material according to the program security classification guide. Most classified information produced by DoD is derivatively classified.

Chapter 3, Section 3.12.8 provides further guidance regarding original and derivative classification markings.

### 1.7.4    Marking    Requirements    for    Classified    Technical Reports

The basic marking requirements for classified technical reports are as follows:

- Overall classification marking (Top Secret, Secret, or Confidential): required
- Classification source of material: required
- Declassification date of material: required
- Date of origin of material: required
- Office that originated the material: required
- Distribution statement: required
- Downgrading date of material: if available from source document
- Intelligence control marking: if required by source document
- Warning notice: if required by source document
- Export control notice: if required by export documents.

The preceding information should be included on the cover and title page of a classified technical report. Chapter 3 provides specific information regarding these marking requirements.

Chapters 2 to 5 provide further guidance regarding the marking of the various elements in a classified technical report.

# Chapter *2*

# *Organization and Design*

## 2.1    Organization

In practice, the technical report printed on paper remains the conventional and most popular means of preparing defense-related technical reports. The continued use of the print technical report can be attributed, primarily, not only to the reduced costs and complexities associated with producing a print technical report (as opposed to preparing, for example, a multimedia technical report) but also to the familiarity and ease associated with reading a print technical report. Furthermore, reading a print technical report, as opposed to reading an online or non-print technical report, is generally faster, resulting in less eyestrain.

Today, technical reports are routinely prepared by an author using one or a combination of several word processing or desktop publishing software programs. Additionally, an editor may make online revisions and an illustrator may provide computer-generated graphics. Ultimately, however, most modern technical reports are created electronically, printed on paper, and distributed and retained electronically and on paper. The appearance of this type of technical report remains basically the same, regardless if it is printed on paper or if it is viewed electronically. The guidance provided herein regarding the organization of technical reports is based on this scenario.

Technical reports are organized into three sections: the front matter, the body or text, and the back matter. Table 2-1 lists the elements within each section of a defense-related technical report in the order in which they should appear and indicates whether the element is required, conditional, or optional. (Table 2-1 is based, in part, on Table 1 in ANSI/NISO Z39.18-2005[2] and Table 1 in ANSI/NISO Z39.18-1995.[4]) The front matter, body, and back matter are discussed in detail in Chapters 3, 4, and 5, respectively.

**Table 2-1.**  Technical Report Elements

|  | Element | Status |
| --- | --- | --- |
| Front Matter | Front Cover | Required |
|  | Disclaimers or Similar Notices (Inside Front Cover) | Conditional |
|  | Title Page | Required |
|  | Copyright (Back of Title Page) | Conditional |
|  | Report Documentation Page (Standard Form 298) | Required |
|  | Abstract | Required |
|  | Table of Contents | Required |
|  | Lists of Figures and Tables | Conditional |
|  | Foreword | Optional |
|  | Preface (Administrative Information) | Required |
|  | Acknowledgments | Optional |
| Body | Executive Summary | Required |
|  | Introduction | Required |
|  | Methods, Assumptions, and Procedures | Required |
|  | Results and Discussion | Required |
|  | Conclusions | Required |
|  | Recommendations | Required |
|  | List of References | Conditional |

**Table 2-1.** Technical Report Elements (Continued)

|  | Element | Status |
|---|---|---|
|  | Appendixes | Conditional |
|  | Bibliography | Optional |
| Back Matter | List(s) of Symbols, Abbreviations, and Acronyms | Conditional |
|  | Glossary | Optional |
|  | Index | Optional |
|  | Distribution List | Required |
|  | Back Cover | Required |

## 2.2 Design

### 2.2.1 Subordination and Format

Subordination and format are intertwined and are, thus, discussed together. Subordination is indicated "by using headings and subheadings to divide the [technical] report into manageable sections, call attention to main topics, and signal changes in topics."[2] "The physical appearance of a [technical] report, both text and graphics, constitutes format. The goal of any format is to enhance readability and comprehension by providing visual uniformity and a consistent subordination of ideas."[2] A good format produces an attractive document. ANSI/NISO Z39.18-2005 prescribes two types of subordination and format for technical reports: "a progression of fonts"[2] or typographical progression and "a decimal numbering system."[2]

Some format features are common to both types of subordination; however, there are significant differences and, thus, advantages and disadvantages to both. These differences also include the placement of classification markings, which are discussed next. The guidelines presented herein for both formats should be read in their entirety; the same information is not necessarily repeated. The figures in Chapters 3, 4, and 5 show the formats for the two types of subordination as they are applied to the various elements in a technical report. The formats prescribed herein are based, in part, on actual formats now being used in the defense community.

### 2.2.1.1 Classification Markings

The main and, perhaps, only difference between unclassified technical reports and classified technical reports with regard to format relates to classification markings. Usually, an unclassified technical report should not be marked unclassified; however, if it is important to convey to the reader that the material has been examined specifically for classification purposes, the technical report should be marked unclassified. In classified technical reports, individual pages should be marked indicating the highest classification of information appearing on that page or they should be marked with the overall classification of the technical report.

Page classification markings should appear in all-capital letters centered in the page header and page footer in a type larger than any other on the page. Page classification markings should not be abbreviated, except for intelligence control markings and warning notices. Special situations regarding page classification markings are discussed in Section 2.2.3.7.

Headings, paragraphs, figures and tables, figure and table titles, footnotes, and, in some cases, lists and equations are considered portions and should be marked individually. Portion markings should be abbreviated.

The following abbreviations should be placed in parentheses, e.g., (S), and used as standard classification markings on headings, paragraphs, figure and table titles, footnotes, and, where applicable, lists and equations in classified technical reports:

- Top Secret: TS
- Secret: S
- Confidential: C
- Unclassified: U.

Where applicable, the previous standard classification markings should be combined with the following abbreviations, which indicate intelligence control markings and warning notices, e.g., (C–RD):

- Authorized for Release to (name of country or international organization): REL TO (abbreviated name of country or international organization [Chapter 3, Table 3–6])
- Caution—Proprietary Information Involved: PR
- Dissemination and Extraction of Information Controlled by Originator: OC
- Not Releasable to Contractors/Consultants: NC
- Not Releasable to Foreign Nationals: NF
- Critical Nuclear Weapon Design Information: N
- Restricted Data: RD
- Formerly Restricted Data: FRD.

Short forms for certain warning notices and intelligence control markings exist and may be combined with spelled-out classification markings, e.g., SECRET NOFORN. These should be used in figures and tables and page footers. The short forms for the intelligence control markings and warning notices are as follows:

- Authorized for Release to (name of country or international organization): REL TO (abbreviated name of country or international organization [Chapter 3, Table 3–6])
- Caution—Proprietary Information Involved: PROPIN
- Dissemination and Extraction of Information Controlled by Originator: ORCON
- Not Releasable to Contractors/Consultants: NOCONTRACT
- Not Releasable to Foreign Nationals: NOFORN
- Critical Nuclear Weapon Design Information: CNWDI
- Restricted Data: Restricted Data
- Formerly Restricted Data: Formerly Restricted Data.

Intelligence control markings and warning notices are explained in Chapter 3, Sections 3.12.3 and 3.12.4, respectively.

### 2.2.1.2  Typographical Progression

Typographical progression is more stylistic than decimal numbering, employing the use of different styles of type, such as boldface and italic, and placing headings in various locations on the page to indicate subordination. Typographical progression is better suited for short- to medium-length technical reports and those not containing more than four heading levels. Because of the absence of decimal numbers, typographical progression is less conducive to extensive cross-referencing. However, unlike a technical report using a decimal numbering format, a technical report using a typographical progression format is not divided into sections, thus allowing more than one first-level heading to be placed on a page in the main body of the technical report, thereby condensing the technical report.

### 2.2.1.2.1  Headings

ANSI/NISO Z39.18–2005 mentions the use of different size fonts to distinguish "primary headings"[2] from "non-primary headings"[2]; however, this is unnecessary and should be avoided in defense-related technical reports, especially classified technical reports. Larger fonts should be reserved only for the overall classification of the technical report and page classification markings. A suggested typeface and font are prescribed in Section 2.2.3.2.

A typical format for typographical progression headings would be as follows:

a. First-level headings: boldface italic initial capital letters centered between the left and right margins. If a first-level heading is longer than one line, a pyramid style arrangement should be used. Runover lines should be single spaced.

b.  Second-level headings: boldface initial capital letters beginning flush left at the margin with flush left runovers. Runover lines should be single spaced.

c.  Third-level headings: italic initial capital letters beginning flush left at the margin with flush left runovers. Runover lines should be single spaced.

d.  Fourth-level headings: italic initial capital letters followed by a period and an em space and placed at the beginning of a paragraph.

In classified technical reports using a typographical progression format, i.e., unnumbered headings, the classification marking should appear in parentheses after the heading on first-, second-, and third-level headings. The classification should appear before the heading on a fourth-level heading; it applies to the heading and the paragraph. The classification marking should not be italicized in first-, third-, or fourth-level headings; however, boldface is acceptable for first- and second-level headings.

Figure 2-1 shows examples of headings, including a pyramid style arrangement, in an unclassified and classified technical report using a typographical progression format.

## 2.2.1.2.2    Paragraphs

Text is arranged in paragraphs, which consist of sentences and, occasionally, equations. Paragraphs should be single column and should be initially indented ½ inch (3 picas) with flush left runovers. Paragraphs should be aligned flush left at the margin with ragged right edges. (Ragged right edges are easier to read than left-right justified edges.) A continuation of a paragraph, e.g., following a list or an equation, should begin flush left with the margin. Lines should be single spaced.

In classified technical reports, the classification marking should appear in parentheses flush left on the first line of the paragraph and the text should begin at its indented location. Some organizations require that the paragraph classification be repeated in parentheses before the first word on the first line of the paragraph on the following page in a continued paragraph. This enables the reader to determine the classification of a paragraph without referring to the previous page. In such cases, the classification marking appearing before the continued paragraph should be identical to the classification marking appearing before the entire paragraph. With the exception of lists (Section 2.2.1.2.3) and, possibly, equations (Section 2.2.4), paragraphs should not be divided into separate classifications, i.e., at the sentence or

---

*First-Level Heading*

**Second-Level Heading**

*Third-Level Heading*

*Fourth-Level Heading.*    The paragraph begins here.
Runover lines are flush left with the margin.

---

**Figure 2–1a.    Unclassified Technical Report**

---

*This Is an Example of a Pyramid
Style Arrangement of a First-Level Heading*

---

**Figure 2–1b.    Pyramid Style Arrangement
in Unclassified Technical Report**

Markings for instructional purposes only.
This figure is unclassified and approved for public release.

---

*First-Level Heading* (U)

**Second-Level Heading (U)**

*Third-Level Heading* (U)

(U) *Fourth-Level Heading.*    The paragraph begins
here. Runover lines are flush left with the margin.

---

**Figure 2–1c.    Classified Technical Report**

---

*This Is an Example of a Pyramid
Style Arrangement of a First-Level Heading* (U)

---

**Figure 2–1d.    Pyramid Style Arrangement
in Classified Technical Report**

**Figure 2–1.    Examples of Headings Using
Typographical Progression Format**

phrase level. An example of a continued paragraph in a classified technical report is shown in Chapter 4, Figure 4-6 (page 164).

### 2.2.1.2.3    Lists

Lists may appear within paragraphs or be displayed separately between sub-paragraphs or at the end of a paragraph. Lists within paragraphs should generally not contain more than five short items; displayed lists should be used in all other instances. Lists within paragraphs should not be identified and separated by number, letter, or other character, e.g., bullet, or contain secondary lists. Lists should be parallel in structure. Errors in parallelism occur most often when nouns and verbs are used inconsistently as the first or principal word in each listed item. For example, a list should not consist of a series of items that contains verbs and include an item that does not contain a verb and vice versa.

Displayed alphanumeric lists in technical reports using a typographical progression format should ideally not contain more than two levels; three levels may be used in rare instances. A bullet list may be combined with a secondary list (using en dashes) but should not be combined with an alphanumeric list and vice versa. Alphanumeric lists should be used to show a particular sequence or to show subordination, e.g., item "a" occurs before item "b" or item "a" is more important than item "b"; bullet/en dash lists should be used when the listed items are of equal importance and their arrangement is random.

Displayed lists should be punctuated as follows: If one or more of the listed items constitute a sentence, a period should be placed after each listed item for consistency in appearance. If one or more of the listed items are phrases but not sentences, the listed items should be punctuated with commas and semicolons, as necessary. The word "and" should appear after the penultimate listing, and the last item should be followed by a period. If the list comprises only short, random items, no punctuation is necessary after each item; however, a period should be placed after the last item if the list is inclusive. The first word in each displayed list item is usually capitalized, regardless if that item constitutes a sentence.

In classified technical reports, lists within paragraphs are the same classification as the paragraph itself. Displayed lists in classified technical reports should be marked individually if one or more of the listed items constitutes a complete thought, i.e., an independent clause or a sentence, regardless if the listed items are the same classification of the paragraph they divide or follow; otherwise, they should not be marked and they are assumed to be the same classification of the paragraph they divide or follow. If a series of listed items does not individually constitute a complete thought but one or more of the listed items are a different classification than the paragraph they divide or follow, they should be marked individually. The classification marking appears in parentheses after the number, letter, or other character but before the text.

A subparagraph following a list of unmarked items should not be marked and should appear flush left with the margin. The subparagraph is assumed to be the same classification as the rest of the paragraph; otherwise, a new paragraph should be created and marked separately.

A subparagraph in a classified technical report following a list of individually marked items should be marked separately. The classification marking appears in parentheses flush left of the subparagraph; the subparagraph is indented slightly to allow space for the classification mark.

Figure 2-2 shows examples of the progression of an alphanumeric first-, second-, and third-level list in a technical report using a typographical progression format. Figure 2-2 also shows examples of the progression of a combined bullet and en dash list in a technical report using a typographical progression format. Further examples of an unmarked list and a marked list and marked subparagraph in a classified technical report using a typographical progression format are shown in Chapter 4, Figure 4-6 (pages 164 and 166, respectively).

### 2.2.1.3    Decimal Numbering

The decimal numbering format is better suited for medium-length to long technical reports and to maintain general consistency in appearance with other military publications. In technical reports using a decimal numbering format, the main body is divided into sections. By compartmentalizing technical reports, the "ripple effect" of changes is reduced; cross-referencing is facilitated by the use of decimal numbers in headings.

MIL-STD-38784,[6] is, with certain exceptions noted herein, a useful guide for establishing a decimal numbering system in technical reports. Section 4.2.9.5.2, "Decimal paragraph numbering,"[6] and Figure 7, "Example decimal paragraph numbering,"[6] of MIL-STD-38784 outline the decimal numbering system.

### 2.2.1.3.1    Headings

Because headings are numbered in technical reports using a decimal numbering format, subordination is apparent. Section headings should be centered, but there is no need to use italic type or place subordinate headings in different locations on the page. Headings may extend to five levels in technical reports using a decimal numbering format because of their as-

1.  First level
    a.  Second level
        (1)   Third level

Figure 2–2a.   Alphanumeric List

• Bullet
    –   En dash

Figure 2–2b.   Combined Bullet and En Dash List

Markings for instructional purposes only.
This figure is unclassified and approved for public release.

1.  (U) First level
    a.  (U) Second level
        (1)   (U) Third level

Figure 2–2c.   Individually Marked Alphanumeric
List in Classified Technical Report

• (U) Bullet
    –   (U) En dash

Figure 2–2d.   Individually Marked Combined Bullet
and En Dash List in Classified Technical Report

Figure 2–2.   Examples of Lists Using
Typographical Progression Format

sumed greater size and complexity. All headings should be boldface initial capital letters to distinguish headings from paragraphs. (The use of all-capital letters and underlining in headings prescribed by MIL-STD-38784[6] for technical manuals is unnecessary in technical reports and should be avoided. This is discussed further in Section 2.2.3.2.) A first-level heading should appear under the section number and should be centered on the page. If a first-level heading is longer than one line, a pyramid style arrangement (Section 2.2.1.2.1) should be used.

Second- through fifth-level headings should begin flush left at the margin with runovers aligned under the first word in the heading. Runover lines should be single spaced for all headings. Figure 2-3 shows the format for decimal-numbered headings.

In classified technical reports using a decimal numbering format, i.e., numbered headings, the classification marking should appear in parentheses after the decimal number but before the heading. In an unnumbered heading in a classified technical report using a decimal numbering format, e.g., abstract and executive summary, the classification marking should appear after the heading as it would in a technical report using a typographical progression format.

Additional information regarding headings is provided in Chapter 3, Section 3.17.

---

**Section 1**
**First-Level Heading**

| | |
|---|---|
| **1.1** | **Second-Level Heading** |
| **1.1.1** | **Third-Level Heading** |
| **1.1.1.1** | **Fourth-Level Heading** |
| **1.1.1.1.1** | **Fifth-Level Heading with Runover Aligned Under First Word in Heading** |

---

**Figure 2-3a.   Unclassified Technical Report**

Markings for instructional purposes only.
This figure is unclassified and approved for public release.

---

**Section 1**
**(U) First-Level Heading**

| | |
|---|---|
| **1.1** | **(U) Second-Level Heading** |
| **1.1.1** | **(U) Third-Level Heading** |
| **1.1.1.1** | **(U) Fourth-Level Heading** |
| **1.1.1.1.1** | **(U) Fifth-Level Heading with Runover Aligned Under First Word in Heading** |

---

**Figure 2-3b.   Classified Technical Report**

**Figure 2-3.   Examples of Headings Using Decimal Numbering Format**

## 2.2.1.3.2    Paragraphs

Paragraphs in technical reports using a decimal numbering format should not be numbered; they are identical to paragraphs in technical reports using a typographical progression format (Section 2.2.1.2.2).

## 2.2.1.3.3    Lists

As with headings, because of the assumed greater size and complexity of technical reports using a decimal numbering format, displayed alphanumeric lists may contain up to five levels. Again, a bullet list may be combined with an en dash list but should not be combined with an alphanumeric list and vice versa.

Figure 2-4 shows examples of the progression of a first- through fifth-level displayed list in a technical report using a decimal numbering format. Figure 2-4 also shows examples of the progression of a combined bullet and en dash displayed list in a technical report using a decimal numbering format.

Displayed lists in a classified technical report using a decimal numbering format should be marked the same as displayed lists in a technical report using a typographical progression format (Section 2.2.1.2.3).

## 2.2.2    Visuals

## 2.2.2.1    Description

Well-prepared technical reports usually contain a balanced arrangement of text and visuals. Visuals comprise figures and tables; they do not include equations, although equations may appear in figures and tables. Figures include charts, graphs, illustrations, maps, and photographs; tables are basically data placed in cells, which are arranged in columns and rows. (Figures and tables are discussed separately in Sections 2.2.2.5 and 2.2.2.6, respectively.)

Figures and tables do not repeat the text; instead, they are a complementary, often clarifying means of presenting information. Figures and tables may appear in the executive summary; the main body, exclusive of the list of references; and in appendixes. Figures and tables should not appear in the abstract. Figures and tables should be separated into two categories: essential and non-essential. Figures and tables essential to the understanding of the technical report should appear in the body of the technical report, and some of these may appear in the executive summary; figures and tables

---

a. First level
    (1) Second level
        (a) Third level
           <u>1</u> Fourth level
              <u>a</u> Fifth level

---

**Figure 2-4a. Alphanumeric List**

---

• Bullet
   – En dash

---

**Figure 2-4b. Combined Bullet and En Dash List**

Markings for instructional purposes only.
This figure is unclassified and approved for public release.

---

a. (U) First level
    (1) (U) Second level
        (a) (U) Third level
           <u>1</u> (U) Fourth level
              <u>a</u> (U) Fifth level

---

**Figure 2-4c. Individually Marked Alphanumeric
List in Classified Technical Report**

---

• (U) Bullet
   – (U) En dash

---

**Figure 2-4d. Individually Marked Combined Bullet
and En Dash List in Classified Technical Report**

**Figure 2-4. Examples of Lists Using Decimal
Numbering Format**

not essential to the understanding of the technical report should appear in an appendix, if at all.

All figures and tables should be mentioned in the text and placed near but not before their first mention. Figures typically "show," and tables typically "list." Other longer, more elaborate verbs, e.g., "illustrate," "depict," "highlight," etc., often used to describe figures and tables are usually unnecessary and should be avoided. If possible, figures and tables should be placed on the same page as their first mention or on the following page if they are first mentioned at the bottom of a page or if they occupy an entire page. However, in some instances, figures and tables may be grouped at the end of an element or section if they are related and mentioned together and their placement near their first mention unduly separates the text and interrupts the flow of the technical report. They should, however, not be placed in another element or section in the technical report. For example, a figure or table mentioned in the methods, assumptions, and procedures should not appear in the results.

Notwithstanding, in extremely short technical reports using a typographical progression format, all figures and tables may be placed at the end of the main body of the technical report; however, they should not appear after the list of references.

Figures and tables are occasionally used in more than one technical report, and this should be considered when creating them. A figure or table is standalone in that aspect and should be able to be understood in its entirety without referring to other parts of the technical report, including the list(s) of symbols, abbreviations, and acronyms. Therefore, all uncommon symbols and abbreviations, including acronyms, in a figure or table should be identified, regardless if they have been identified previously in the technical report.

No new figures or tables should appear in the executive summary; basically, figures and tables appearing in the executive summary are repeated from the main body of the technical report, although they may be modified. Figures and tables appearing in the executive summary and/or main body should not be repeated in an appendix, and appendix figures and tables should not be repeated in the executive summary and/or main body.

ANSI/NISO Z39.18-2005[2] states that lowercase superscript letters should be used in footnotes to figures and tables; however, to avoid confusion with text footnotes that use the same system, lowercase Roman numerals may be used if there are many footnotes, or a sequence of symbols, such as an asterisk (*), dagger (†), double dagger (‡), etc., may be used if there only a few footnotes.

Figures and tables should not divide a paragraph. Large figures and tables occupying more than half of the page should be placed at the top or bottom of a page; they should not be "sandwiched" in the middle of a page between text. Conversely, text should not be inserted between two large figures or tables. The text should either precede or follow a figure or table. Separate figures and tables may be stacked on a page but should not be placed side by side (with the exception of subfigures). Figures and tables should, if possible, be oriented vertically or in a portrait mode on an 8½-inch-by-11-inch page. However, there are instances where this is not feasible and the figure or table is required to be oriented horizontally or in a landscape mode, i.e., rotated counterclockwise 90 degrees, on an 8½-inch-by-11-inch page.

Additional information regarding figures and tables is provided in Chapter 3, Section 3.18.

## 2.2.2.2 Numbering and Titling

All figures and tables should be numbered and titled, and they should be mentioned by number in the text, not by "the following figure" or "the preceding table." The only exception to this is in a technical report using a typographical progression format and the technical report contains only one figure or table; then the figure or table is unnumbered but is titled. Here, the figure or table is mentioned in the text as, simply, "the figure" or "the table." All figures and tables in a technical report using a decimal numbering format are numbered and titled, regardless of the number.

Figures and tables are numbered separately using an identical numbering scheme. In a technical report using a typographical progression format, the numbering scheme for figures and tables does not begin in the executive summary and continue to the main body. Instead, figures and tables in an executive summary should be numbered ES-1, ES-2, etc. In the main body, they should be numbered 1, 2, 3, etc. If a figure or table is repeated in the executive summary from the main body, it is renumbered as an executive summary figure or table. In an appendix, figures and tables are numbered A-1, A-2, B-1, B-2, etc., with the letter referring to the appendix, unless the appendix is a previously published document, whereby the original numbering scheme is maintained and A-1, A-2, etc., are added as secondary numbers, e.g., Figure 1/A-1.

In a technical report using a decimal numbering format, figures and tables in an executive summary are also numbered ES-1, ES-2, etc. In the main body they are numbered by section, e.g., 1-1, 1-2, 2-1, 2-2, etc., with the first number referring to the section number and the second number referring to the figure or table number. The numbering scheme restarts with each new section. In an appendix, they are also numbered A-1, A-2, B-1, B-2, etc. The other aspects of numbering are the same as they are in technical report using a typographical progression format.

Figure or table titles should be informative and concise using a specification or telegraphic style of wording, which condenses writing by deleting superfluous articles ("a," "an," and "the") without losing comprehension. If comprehension is affected, these articles should be left in the title. The complete title should begin with the word "Figure" or "Table," followed by the appropriate number, a period, and en space.

Styles vary for figure or table titles. In one style (used herein), all words, except articles and prepositions containing four or less letters, are capitalized. No punctuation is placed at the end of the title; however, ancillary sentences accompanying the title are placed in parentheses after the title and punctuated accordingly (Figure 2-15a and Table 2-16a). In another style, the first word is capitalized and the remaining words are lowercase unless they are a proper noun or adjective. A period is placed at the end of the title.

Figure or table titles extending more than one line should be flush left or arranged in an inverted pyramid style (used herein). Runover lines are single spaced. Figure or table titles should be typed in a sans serif typeface, e.g., Helvetica, to distinguish them from the text. The word "Figure" or "Table" and number and period following it should be in boldface type; the rest of the title should be in lightface type.

Occasionally, figures and tables will comprise two or more subfigures or subtables. In these instances, the overall figure or table is numbered and titled as usual and each subfigure or subtable is numbered and titled as follows: A lowercase letter is placed after the figure number, e.g., Figure 1a, Figure 1b, Table 1-1a, Table 1-1b, etc., and each subfigure and subtable is given a unique subtitle.

## 2.2.2.3    Classified Technical Reports

In classified technical reports, there are two separate classifications for a figure or table: The figure or table, itself, and the figure or table title are classified separately, and each may have a different classification. The classification of the figure or table is spelled out and usually appears under the lower right-hand corner of the figure or table. This classification applies to the figure or table in its entirety, including footnotes to the figure or table. If a figure or table continues beyond one page, each portion of the figure or table is marked. This occurs frequently with tables. Separate or additional classifications should not appear within a figure or table. If applicable, the short form of an intelligence control marking (Section 2.2.1.1) is added to the figure or table classification. The classification for a figure or table title is abbreviated and appears after the figure or table number but before the title. Figure or table subtitles are treated the same as figure or table titles.

## 2.2.2.4    Examples

The examples in Figures 2-5 to 2-12 show figures in a typographical progression format; however, they are equally applicable to tables and a decimal numbering format. Figures 2-5 and 2-6 show examples of a portrait and landscape orientation in an unclassified and classified technical report, respectively. The page header and footer in a landscape orientation should remain in the same location as in a portrait orientation; this is often referred to as a "turnpage."

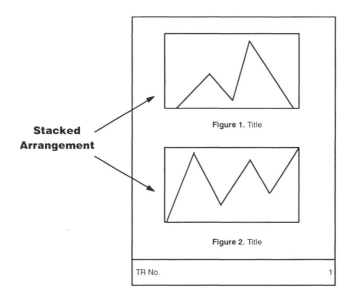

**Portrait Orientation**

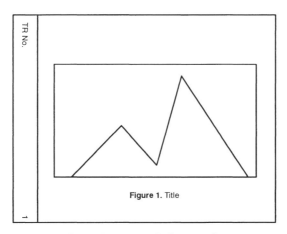

**Landscape Orientation**

Figure 2-5.   Examples of Portrait and Landscape
Orientations in Unclassified Technical Report

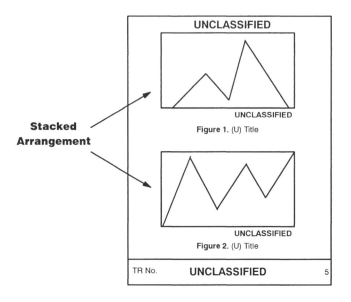

**Portrait Orientation**

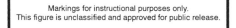

Markings for instructional purposes only.
This figure is unclassified and approved for public release.

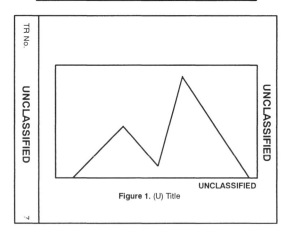

**Landscape Orientation**

**Figure 2-6.   Examples of Portrait and Landscape
Orientations in Classified Technical Report**

Usually, a single figure occupies not more than one page. Facing pages are used in instances where a figure or table is too large to fit onto an 8½-inch-by-11-inch page and a foldout is not feasible, possibly because of publishing limitations. Facing pages are also used when two or more figures or two or more tables are closely related and it is important that they be viewed together. A facing page begins on a new even-numbered, left-hand page and continues to an odd-numbered, right-hand page, thereby allowing it to be viewed in its entirety without turning a page. Facing pages should only be used in portrait orientations. If a facing page is used for one figure or one table, the transition from the first page to the second page should be "seamless," despite the binder separation, without loss of data. Figures 2-7 and 2-8 show examples of subfigures and facing pages for a single figure in an unclassified and classified technical report, respectively.

There are also instances where a figure or table cannot fit onto an 8½-inch-by-11-inch page and facing pages are impractical; here, a foldout should be used. A foldout is always a new odd-numbered, right-hand page; therefore, the technical report number should appear on the lower left-hand corner and the page number should appear on the lower right-hand corner. The figure or table title should appear on the top fold. The figure title should appear at the bottom of the fold; the table title should appear at the top of the fold. In a classified technical report, the page classification should appear on the top and bottom center of the top fold so it is evident without unfolding the page. The classification of the figure or table and the classification of the figure or table title should appear as they normally would. The back of a foldout should be blank in unclassified and classified technical reports but should serve as the following even-numbered page in the technical report.

Figures 2-9 and 2-10 show the correct method of creating a 17-inch-by-11-inch foldout and a 22-inch-by-11-inch foldout in an unclassified technical report, respectively; Figures 2-11 and 2-12 show the correct method of creating a 17-inch-by-11-inch foldout and a 22-inch-by-11-inch foldout in a classified technical report, respectively.

## 2.2.2.5    Figures

The importance of good figures cannot be overstated. If used correctly, figures, as opposed to text, can more easily explain a complicated subject and turn an otherwise ordinary technical report into an interesting and, even, entertaining document. Conversely, bad figures usually result in a bad technical report. Figures 2-13[5] and 2-14[5] show the importance of figures in absorbing and retaining information and the relationship between text and figures.

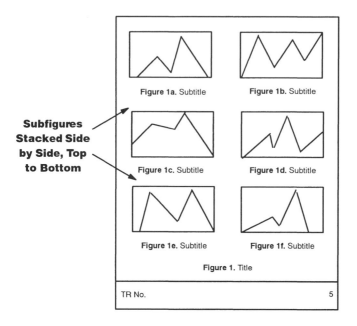

**Subfigures Stacked Side by Side, Top to Bottom**

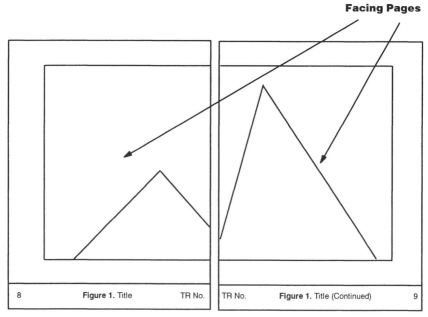

Figure 2-7.    Examples of Subfigures and Facing
Pages in Unclassified Technical Report

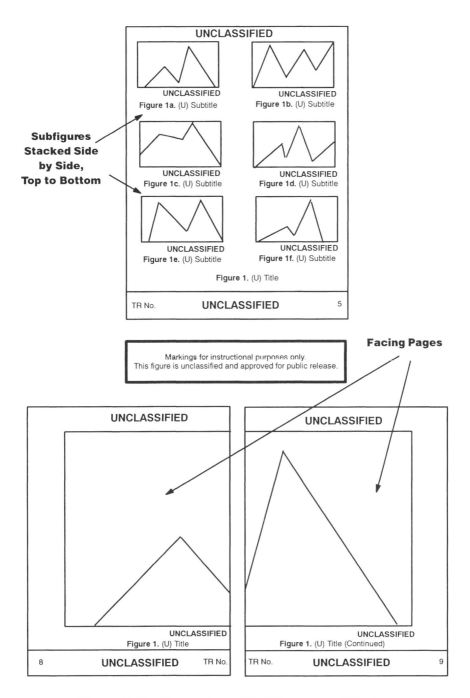

Figure 2-8.    Examples of Subfigures and Facing
Pages in Classified Technical Report

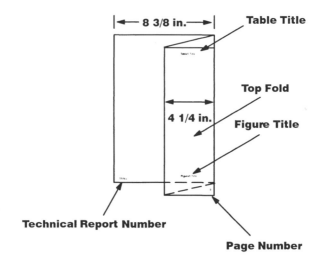

Figure 2-9.    Example of 17-Inch-by-11-Inch Foldout
in Unclassified Technical Report

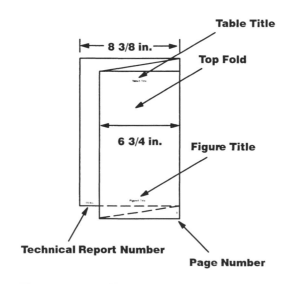

Figure 2-10.    Example of 22-Inch-by-11-Inch
Foldout in Unclassified Technical Report

Figures are many and varied; however, they can be grouped into categories. ANSI/NISO Z39.18-2005 draws the following distinctions for most
figure types:

The type of figure used depends on the type of information being presented: graphs show relationships among data; diagrams portray rela-

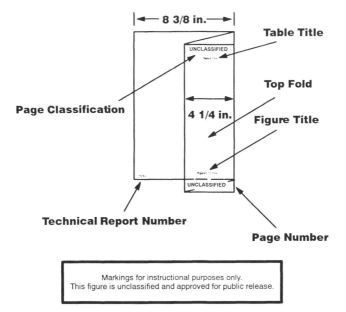

Figure 2–11.    Example of 17-Inch-by-11-Inch
Foldout in Classified Technical Report

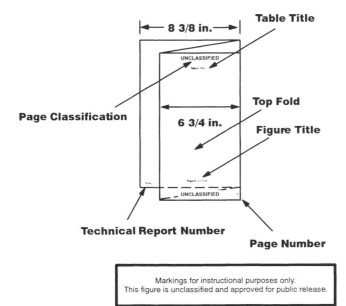

Figure 2–12.    Example of 22-Inch-by-11-Inch
Foldout in Classified Technical Report

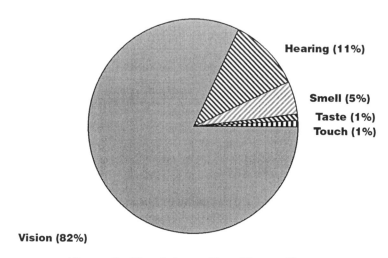

Vision (82%)

Figure 2–13.    Information Absorption

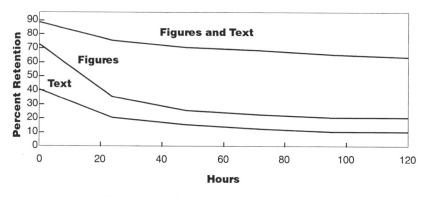

Figure 2–14.    Information Retention

tionships among components; photographs realistically depict general appearance; and drawings emphasize essential elements and omit unnecessary details.[2]

While figures may be graphs, diagrams, photographs, or drawings, they should not be labeled as any of these subcategories; they are always "figures." Some charts and graphs may be tabular in nature; however, they, too, should be labeled as figures.

Charts and graphs are similar. There are various types of charts, including bar charts, organizational charts, pie charts, and flow charts. Graphs show trends in data. Graphs usually contain an "X" or horizontal axis and a "Y" or vertical axis. Y-axis callout letters should appear in a landscape mode (Figure 2-14), not stacked top to bottom. Illustrations and diagrams are synonymous and encompass drawings; they should be used when a photograph is impractical or unnecessary. In addition, maps, like charts, are diverse; they are usually topographical in nature.

Figures may be "boxed" or have a border. Borders are usually unnecessary; however, they are sometimes used to distinguish items within the figure, e.g., subfigures. If used, borders should be confined to the figure or subfigure itself and not include the figure or subfigure title. Figures should not be cluttered. A common mistake of many authors is to include too much information in a single figure. ANSI/NISO Z39.18-2005 states the following:

> Normally, a figure should emphasize one main idea and show no more than is necessary. Figures should have informative titles (captions) that summarize the figure and, as needed, callouts that clearly and concisely identify each part. The figure number and title should appear below the figure. The title describes the content without giving background information, results, or comments about the figure. The placement and alignment of callouts should be consistent within the [technical] report.[2]

The use of color used to be restricted in figures because of the increased publishing costs and the effect certain colors, such as red, had on the printed page—known as "bleeding." However, while publishing in color remains costly, other factors have mitigated over time through advances in publishing and color is being used more frequently. Notwithstanding, much like the overuse of different typefaces in text, the use of color can be overdone and is often unnecessary. Creative ways of using black-and-white alternatives, e.g., shades of gray, patterned lines, etc., should be investigated before using color (Figure 2-13).

Oral presentation materials such as viewgraph compilations are increasingly being used as a foundation for defense-related technical reports. While these materials are frequently annotated, thus containing text, their original purpose is different than a conventional technical report, which is meant to be perused and absorbed, not quickly seen and heard at a "snapshot" pace. As such, these materials should be modified to conform to accepted practices for figures in technical reports and not left unaltered in their original state.

Callouts and footnotes in figures should be typed in a sans serif typeface, e.g., Helvetica (10 points maximum; 6 points minimum). Figure 2-15 shows the common components and classification markings of a figure in an unclassified and classified technical report.

MIL-STD-38784,[6] which provides guidance for preparing illustrations in technical manuals, may also be used as a guide for preparing figures in defense-related technical reports.

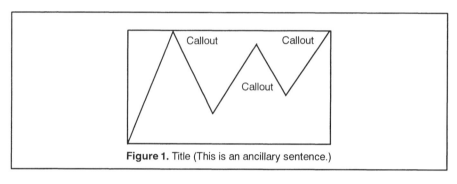

Figure 1. Title (This is an ancillary sentence.)

**Figure 2-15a.    Figure in Unclassified Technical Report**

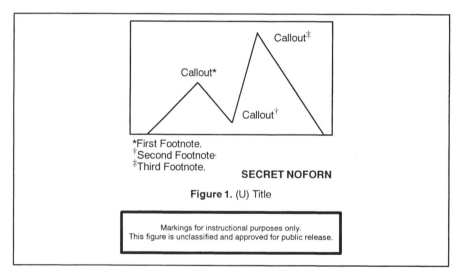

**Figure 2-15b.    Figure in Classified Technical Report**

**Figure 2-15.    Figure Components and Classification Markings**

## 2.2.2.6    Tables

Tables, like figures, are essential to presenting data that would be difficult to present textually. As stated previously, tables comprise columns, rows, and cells. Tables should normally contain at least three columns and two rows. If a table contains only two columns, a list should be considered as an alternative; if a table contains only one row, the information should be presented in the text. Tables also contain a header row and, possibly, a subheader row. Within the table header and subheader rows are headings that apply to particular columns. A table may also contain a footer, which contains footnote information. If a table occupies less than two pages, it should, if possible, appear on a single page and should not be divided.

If a unit of measurement applies to all data in a column, the unit of measurement symbol or its abbreviation, e.g., %, lb, etc., should be placed in parentheses after the column heading; it should not be repeated after each entry in the column. If all data in a particular column are identical, that column is usually unnecessary and should be deleted and the information should be presented in the text at the first mention of the table.

Footnote indicators in tables should appear left to right, top to bottom. The data in a column should, if possible, be centered and aligned. If the data within a cell occupy more than one line, the data should be flush left with the cell and the runovers should be single spaced. Data being totaled in a column should be aligned flush right for whole numbers. Data containing decimals should be aligned with the decimal point, and data containing commas should be aligned with the comma. An en dash should be inserted in blank cells in a table to indicate that data have not been inadvertently omitted.

Borders are contiguous lines or rules surrounding a table. Vertical rules are used to separate data in columns; horizontal rules are used to separate data in rows. Borders and rules should be kept to a minimum; they are usually unnecessary and may, in fact, hinder comprehension. Horizontal rules are only required to separate a header and footer from the rest of the table. Occasionally, table cells are shaded to distinguish certain data. Like figures, shades of gray should be considered before using colors.

The table number and title should be placed above the table. Tables, unlike figures, can often continue for more than one page; if they do, the title should be repeated and "Continued" or "Cont." should be placed after the title in parentheses. The header and footer should be repeated, as well; however, it is unnecessary to insert "Continued" or "Cont." in the header and footer. The table should be continuous and should not be interrupted by text.

Headings, data, and footnotes in tables should be typed in a sans serif typeface, e.g., Helvetica (10 points maximum; 6 points minimum). Figure 2–16 shows the common components and classification markings for a table in an unclassified technical report and a classified technical report.

### 2.2.3    Page Format

Page format consists of the following:
- Column format and line length
- Typography
- Image area
- Margins
- Paper and ink
- Pagination.

Publishing has its own units of measurement: picas and points. Six picas roughly equal 1 inch, and 12 points equal 1 pica. Line length, image area, and margins are measured in picas. Typography is measured in points. Unless indicated otherwise, the page format is the same for technical reports using a typographical progression format and a decimal numbering format.

In classified technical reports, page format is not only affected by classification markings but also by special situations, i.e., compilation of information (Section 2.2.3.7), and blank pages (Section 2.2.3.8).

### 2.2.3.1    Column Format and Line Length

ANSI/NISO Z39.18–2005[2] prescribes a single-column format and a double-column format for technical reports; however, a double-column format for defense-related technical reports is unusual and is usually reserved for documents other than technical reports, such as military specifications and standards and technical manuals. Thus, defense-related technical reports should be in a single-column format. Line length corresponds to type: The larger the type, the longer the line. Notwithstanding, the trend is toward shorter lines. ANSI/NISO Z39.18–2005 prescribes a standard line length of 40 to 43 picas and a minimum line length of 38 picas. This length might be excessive for a smaller type. As an alternative, a standard line length of 33 picas with a minimum line length of 31 picas may be used for 10- to 12-point type.

### 2.2.3.2    Typography

Typography or typeface is often confused with font. Typography includes a number of styles; however, with the exception of mathematical and other

**Table 1.** Title (This is an ancillary sentence.)

| Heading | | Heading | |
|---|---|---|---|
| Subheading | Subheading | Subheading (%) | Subheading |
| Data | Data | 10 | Data |
| Data | Data | 30 | Data |
| Data | Data | 20 | Data |
| Data | Data | 15 | Data |
| Data | Data | 25 | Data |
| | Total | 100 | |

**Data Being Totaled**

**Figure 2-16a.    Table in Unclassified Technical
Report**

**Table 1.** (U) Title

| Heading | Heading | Heading |
|---|---|---|
| Data* | Data | Data |
| Data | Data$^{\dagger}$ | Data |
| Data | Data | Data |
| Data | Data | Data$^{\ddagger}$ |
| Data | Data | Data |

*First Footnote.
$^{\dagger}$Second Footnote.
$^{\ddagger}$Third Footnote.

**SECRET NOFORN**

Markings for instructional purposes only.
This figure is unclassified and approved for public release.

**Figure 2-16b.    Table in Classified Technical Report**

**Figure 2-16.    Table Components and Classification
Markings**

symbols, only serif and sans serif typefaces, combined with other styles, such as boldface, lightface, roman, and italic, apply to technical reports. Times is an example of a serif typeface; Helvetica is an example of a sans serif typeface. A font comprises one size (measured in points) and style within a typeface.

Times or a comparable serif typeface should be used for the text, including headings; Helvetica or a comparable sans serif typeface should be used in figures and tables. ANSI/NISO Z39.18-2005[2] advocates a 10- to 12-point serif typeface for the text. (The text herein is 11-point Times and is suitable for defense-related technical reports.) Figure and table titles may be typed in a sans serif 10-point typeface (used herein); however, the titles should not be larger than the text. Callouts in figures and tables should, ideally, be typed in a sans serif 8- to 10-point typeface; however, no callouts should be smaller than 6 points. Visual-only text in figures not intended to be read may be smaller than 6 points.

The use of all-capital letters and underlining is, in many ways, a carry-over from the typewriter era and, with the exception of certain listed items (Section 2.2.1.3.3), is not advocated herein, even for technical reports using a decimal numbering format. While all-capital letters and underlining are still being used as distinguishing features in headings in certain military documents, e.g., military handbooks, their continued use has been reduced if not eliminated by advances in publishing and the ability to modify text by using boldface and italic types or combinations thereof. Furthermore, the frequent and repetitive use of acronyms in defense-related technical reports discourages if not precludes the use of all-capital letters for ship names, weapons systems, etc.

### 2.2.3.3    Image Area

The space allotted on a page . . . for textual, visual, or tabular matter is the image area. Observing a standard image area ensures the information on a page will not be lost during printing and binding. The normal image area on U.S. standard paper that is 8-1/2 by 11 inches . . . is 7-1/8 inches [about 43 picas] by 9-3/16 inches [about 55 picas]. The image area includes headers and footers, if used, and page numbers.[2]

The effective use of white space is an important consideration in establishing an image area. Thus, for a larger white space and a smaller image area, a typical maximum image area would be 33 picas (wide) for text and 39 picas (wide) for figures and tables on a portrait page and 49 picas (wide) for figures and tables on a landscape page. Text would be indented 3 picas from the inner and outer margins (6 picas total). Text should not appear on a landscape page, only figures and tables. In classified technical reports, the page classification markings should appear within the image area.

### 2.2.3.4    Margins

Margins are related to the image area and include page headers and footers. The bottom margin is wider than the top margin. The inner margin may be wider than the outer margin; however, they should be the same if binding is unaffected. In a typical format, the bottom margin contains a 1-point rule separating the text from the technical report number and the page number and, in a classified technical report, the page classification marking. The top margin should be blank unless the technical report is classified; if so, the page classification marking should be inserted in the top margin.

The following page margins are typical on an 8½-inch-by-11-inch page:

- Top margin: 6.5 picas
- Bottom margin: 8.5 picas
- Inner and outer margins: 6 picas.

Figures 2–17 and 2–18 show typical image areas and margins for an unclassified and classified technical report, respectively.

### 2.2.3.5    Page Numbers

Page numbers should appear in alternating locations in defense-related technical reports. On odd-numbered, right-hand pages, page numbers should appear on the right edge of the footer; on even-numbered, left-hand pages, they should appear on the left edge of the footer. By doing so, the page number always appears on the outside edge of the page on technical reports printed back to back or on both sides. Also, this format allows the page classification to appear in the bottom center of the footer in classified technical reports, which is a requirement.

The front matter is numbered with lowercase Roman numerals, and the executive summary is numbered ES–1, ES–2, etc., in technical reports using a decimal numbering format. In technical reports using a typographical progression format, the pages are numbered 1, 2, 3, etc., beginning with the executive summary and continuing to the end of the technical report. In technical reports using a decimal numbering format, each section of the technical report is numbered 1–1, 1–2, 2–1, 2–2, etc., with the first number referring to the section number and the second number referring to the page number in the section. The pagination restarts with each section. Appendixes are paginated A–1, A–2, B–1, B–2, etc., in technical reports using a decimal numbering format. The other elements of the back matter are pagi-

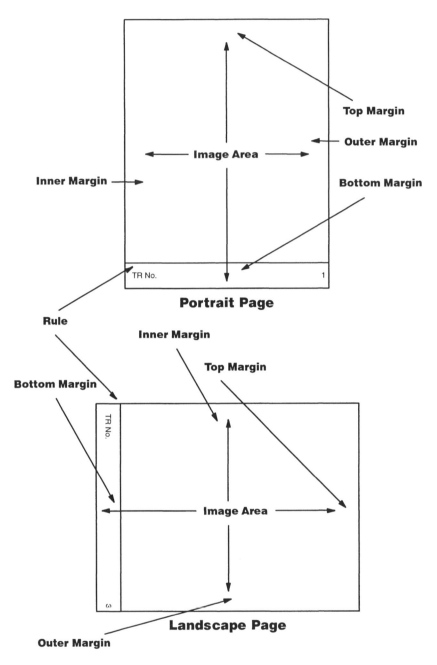

Figure 2-17.    Image Areas and Margins in
Unclassified Technical Report

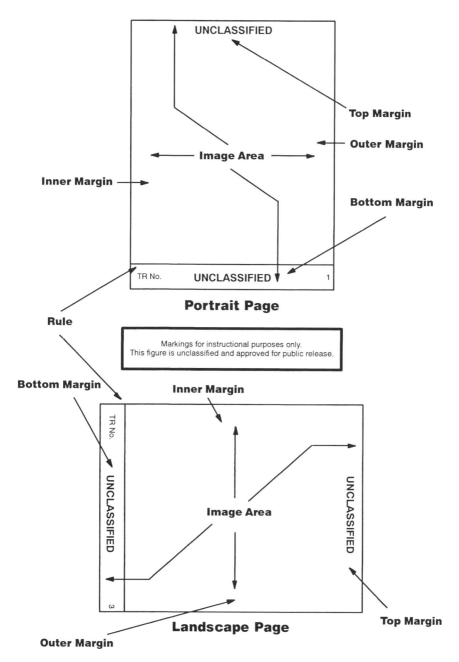

UNCLASSIFIED

Top Margin

Outer Margin

Image Area

Inner Margin

Bottom Margin

TR No.    UNCLASSIFIED    1

**Portrait Page**

Rule

Markings for instructional purposes only.
This figure is unclassified and approved for public release.

Bottom Margin

Inner Margin

TR No.

UNCLASSIFIED

Image Area

UNCLASSIFIED

Top Margin

3

**Landscape Page**

Outer Margin

Figure 2-18.   Image Areas and Margins in
Classified Technical Report

nated in a manner whereby the element name or an abbreviation thereof precedes the page number in that element, e.g., SAA-1 (page 1 in the list[s] of symbols, abbreviations, and acronyms), Glossary-1, ID-1 (first page in the initial distribution), etc.

### 2.2.3.6    Printers and Paper and Ink

Most technical reports are printed on a laser printer, which produces camera-ready copy. Laser printers are capable of printing in black and white or in color. Black-and-white printers should be capable of printing 300 dots per inch resolution; color printers should be able to print 600 dots per inch resolution.

Standard white 8½-inch-by-11-inch acid-free, laser-quality paper should be used unless foldouts are required. Some organizations substitute light-colored paper for the abstract and executive summary to distinguish them from the rest of the technical report; however, this is optional and should not be used for an executive summary if it contains color figures or tables. In classified technical reports, the cover is usually a different color for each of the different classifications to make it conspicuous; the color scheme varies from organization to organization.

Black ink is standard; however, other colors may be used in color figures or tables for contrasting effects.

### 2.2.3.7    Compilation of Information in Classified Technical Reports

### 2.2.3.7.1    Unclassified    Portions    Requiring    Overall Classification

Portions of an otherwise unclassified technical report may, when compiled, constitute classified information, thereby creating a classified technical report. When this occurs, each interior page should be marked with the overall classification level; portion marking is unnecessary. A notice (Chapter 3, Section 3.12) with the following information should appear on the front cover and title page explaining the reason for the classification:

- The fact that the individual portions of the technical report are unclassified;
- The reason why the otherwise unclassified compilation requires classification; and
- The authority for the classification.

## 2.2.3.7.2    Classified Portions Requiring Higher Overall Classification

An overall classification level higher than the individual classified portions of a classified technical report may be required. For example, individual Confidential portions may, when compiled, add up to a Secret overall classification level. When this occurs, each interior page should be marked with the highest classification of information on or revealed by the page, and each portion should be marked with the appropriate classification marking. A notice (Chapter 3, Section 3.12) with the following information should appear on the front cover and title page:

- The fact that the individual portions are of a lower classification level than the technical report as a whole;
- The reason why the compilation requires a higher classification level than its individual portions;
- The authority for the classification; and
- The affected pages.

## 2.2.3.8    Blank Pages

Most paper technical reports are two-sided or printed back to back. In these instances, left-hand or otherwise even-numbered pages that contain no information are typically left blank in unclassified and classified technical reports, i.e., they contain no page number or classification marking. The blank page is included in paginating the technical report, and the next page containing information is an odd-numbered, right-hand page.

To alleviate concerns over possible missing classified information, some organizations require that the blank page be noted in classified technical reports. One approach is to include two page numbers on the preceding odd-numbered, right-hand page, e.g., "2-3/2-4 (blank)" or "23/24 (blank)." Alternatively, some organizations insert the following: "(U) This page intentionally left blank." or a similarly worded sentence or phrase on the otherwise blank page. (The page classification, i.e., **UNCLASSIFIED**, or overall classification of the technical report is also included in the latter case.) Either approach is acceptable, but the former is preferred.

## 2.2.4    Equations

ANSI/NISO Z39.18-2005 discusses "formulas and equations"[2]; however, because the two are closely related if not synonymous as they relate to mathematics, the guidance provided herein for equations applies to both. ANSI/NISO Z39.18-2005 also discusses "mathematical equations"[2] and "chemical equations"[2] and prescribes numbering them independently; however, for simplicity, the two may be numbered together in defense-related technical reports.

Other than the correctness of the equation itself, the most important issues with regard to an equation in a technical report are punctuation and layout. In relation to sentence structure, equations are complete thoughts and are, at a minimum, independent clauses and, more often, complete sentences. They should be punctuated accordingly.

Equations may appear within the text or they may be displayed on a separate line(s). Short, less significant equations may appear in the text and do not need to be numbered. If they appear within the text, they are punctuated as any other independent clause or sentence and should be the same height as the rest of the text. To keep an equation the same height size as the rest of the text, signs of aggregation (discussed later in this section) and virgules or forward slashes should be used in lieu of the typical mathematical notation used in equations displayed on a separate line. For example,

$$\frac{2}{a + b} \times c = d$$

would appear as $[2/(a+b)] \times c = d$ in the text. If, after reformatting, an equation still exceeds the height of the text, it should, if possible, be displayed on a separate line, regardless of its length or significance.

Longer, more significant equations should always appear on a separate line either in the center of the page or indented from the left edge. If there are more than a few equations, they should be numbered and mentioned in the text by number. The number appears in parentheses flush right of the equation and is aligned with the last line of the equation. If they are displayed on a separate line(s), certain punctuation marks may be omitted, e.g., ending comma, semicolon, or period, allowing the spacing of the display to replace punctuation.

Groups within equations are enclosed by parentheses, brackets, and braces, otherwise known as signs of aggregation or fences, in the following sequence: parentheses inside brackets inside braces, i.e., $\{[()]\}$. The sequence repeats itself if the equation contains more than three group levels.

The following guidance, derived in part from the *United States Government Printing Office Style Manual 2000*,[20] addresses equations:

a.  In mathematical equations, italic type should be used for all letter symbols representing numbers, i.e., variables and constants. This includes capital letters, lowercase letters, small capital letters, and superiors (superscripts) and inferiors (subscripts and exponents). Roman type should be used for numerals, including superiors and inferiors. Whole or abbreviated words, including single-letter abbreviations, appearing as superscripts or subscripts should be roman type. Vectors should appear in a serif typeface, e.g., Times boldface or boldface italic. Tensors should appear in a sans serif typeface, e.g., Helvetica boldface or Helvetica boldface italic, to distinguish them from vectors. Chemical symbols should be roman type.

b.  If an equation or a mathematical expression needs to be divided, it should be broken before +, -, =, etc. However, the equal sign should clear on the left of other beginning mathematical signs. An equation should not exceed the normal line length for text, e.g., 33 picas, in width or be divided within a group.

c.  A short equation in text should not be broken at the end of a line. The line should be spaced so that the equation will begin on the next line; or better, the equation should be centered on a line by itself.

d.  An equation too long for one line should be set flush left, the second half of the equation should be set flush right, and the two parts should be balanced as nearly as possible.

e.  Two or more equations in a series should be aligned on the equal signs and centered on the longest equation in the group.

f.  Connecting words of explanation, such as "hence," "therefore," and "similarly," should be set flush left either on the same line with a resulting or follow-up equation or on a separate line after an initial equation. Connecting words should be lowercase.

g.  Symbols in equations should be consistent in appearance with corresponding symbols in the text.

h.  Parentheses, braces, brackets, integral signs, and summation signs should be of the same height as the mathematical expressions they include.

i.  Inferiors should precede superiors if they appear together; but if either inferior or superior is too long, the two should be aligned on the left.

In classified technical reports, equations are usually not marked because they are part of a paragraph. Equations embedded in the text should not be marked separately; if, however, the equation is displayed on a separate line(s) and the classification of the equation is different from the rest of the paragraph, the classification marking should appear in parentheses flush left of the first line of the equation.

Equations should appear in the same serif typeface used for the text. Figure 2-19 shows examples of equations in unclassified and classified technical reports using a typographical progression format and a decimal numbering format.

$$E = mc^2 \tag{1}$$

where,

> $E$ is a quantity of energy;
> $m$ is its mass;
> $c$ is the speed of light.

**Figure 2-19a.   Example of Equation in Unclassified Technical Report Using a Typographical Progression Format**

(U)                      $$E = mc^2 \tag{1}$$

**Figure 2-19b.   Example of Equation in Classified Technical Report Using a Typographical Progression Format**

Markings for instructional purposes only.
This figure is unclassified and approved for public release.

$$E = mc^2 \tag{1-1}$$

where,

> $E$ is a quantity of energy;
> $m$ is its mass;
> $c$ is the speed of light.

**Figure 2-19c.   Example of Equation in Unclassified Technical Report Using a Decimal Numbering Format**

(U)                      $$E = mc^2 \tag{1-1}$$

**Figure 2-19d.   Example of Equation in Classified Technical Report Using a Decimal Numbering Format**

**Figure 2-19.   Examples of Equations in Unclassified and Classified Technical Reports**

## 2.2.5    Footnotes

In addition to providing supplementary or clarifying information to the text, footnotes serve an additional purpose in defense-related technical reports, i.e., the identification of unpublished references. Unpublished references should not be included in the list of references; instead, they should be footnoted. (Web sites and documents available on the Internet should be considered "published.") These include memorandums, draft documents, and personal communications, such as letters, telephone conversations, and e-mail.

Footnotes should be kept to a minimum and should not appear in an abstract or an executive summary. Footnotes are numbered consecutively throughout the main body of a technical report and may continue into the appendixes. This applies to both formats. Repeated footnotes should be assigned a new number. Footnotes are often indicated by a superscript Arabic number; however, to avoid confusion with reference citations using the same numbering scheme, a superscript lowercase Roman letter should be substituted. If superscript Arabic numbers are not used to indicate reference citations, they may be used to cite footnotes. If the technical report contains only one footnote, a symbol, e.g., asterisk, should be used in lieu of a letter or number; however, the symbol should be different from any figure or table footnote(s) on the same page. Text footnotes should be numbered separately from figure and table footnotes.

In classified technical reports, a footnote is classified separately from the text it refers to. The classification marking appears in parentheses to the immediate right of the superscript letter or number and to the left of the text it refers to.

Footnotes should appear in a smaller type, e.g., 9- or 10-point Times, than the text at the bottom of the page on which they appear (above the footer). Typically, a 1-inch horizontal rule is placed above the topmost footnote on the page, flush left with the margin. Footnotes are single spaced. Figure 2-20 shows examples of footnotes in unclassified and classified technical reports.

## 2.2.6    Errata

Published and distributed technical reports occasionally contain technical errors. A "technical error" should be significant and should not include matters of format and style or grammar and punctuation unless they affect the meaning of the technical report. A technical error should be noted in an errata sheet. The errata sheet is distributed to those on the primary distribution list. In technical reports using a typographical progression format, a text error should by identified by page and line number; in technical reports

---
ᵃText.
---

**Figure 2–20a.   Example of Footnote in Unclassified Technical Report**

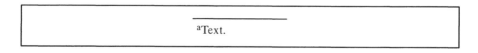

Markings for instructional purposes only.
This figure is unclassified and approved for public release.

---
ᵃ(U) Text.
---

**Figure 2–20b.   Example of Footnote in Classified Technical Report**

**Figure 2–20.   Examples of Footnotes in Unclassified and Classified Technical Reports**

using a decimal numbering format, a text error should be identified by section number and line number. Figures and tables should be identified by their number.

In classified technical reports, the classification of erroneous information and the correct information should be noted by placing a classification marking to the left of both. The classification of the errata sheet should be identified in the form of a page classification. Further instructions should be provided if the paragraph, figure, or table classification changes.

Figure 2–21 shows an errata sheet for an unclassified technical report and a classified technical report using either format.

| Page | Reads | Should Read |
|------|-------|-------------|
| 5, Line 8 | _____ | _____ |

**Figure 2-21a.   Errata Sheet in Unclassified
Technical Report Using Typographical Progression
Format**

| Section | Reads | Should Read |
|---------|-------|-------------|
| 3.5, Line 3 | _____ | _____ |

**Figure 2-21b.   Errata Sheet in Unclassified
Technical Report Using Decimal Numbering Format**

Markings for instructional purposes only.
This figure is unclassified and approved for public release.

| Page | Reads | Should Read |
|------|-------|-------------|
| 5, Line 8 | (U)_____ | (C)_____ |

**Figure 2-21c.   Errata Sheet in Classified Technical
Report Using Typographical Progression Format**

| Section | Reads | Should Read |
|---------|-------|-------------|
| 3.5, Line 3 | (S)_____ | (S)_____ |

**Figure 2-21d.   Errata Sheet in Classified Technical
Report Using Decimal Numbering Format**

**Figure 2-21.   Errata Sheet Format**

Chapter *3*

# *Front Matter*

The front matter of defense-related technical reports should contain the following elements in the order presented:

    a.  Front cover (including inside front cover) (The back cover, while part of the back matter of a technical report, is discussed in this chapter.)

    b.  Title page (including back of title page)

    c.  Report documentation page (Standard Form 298)

    d.  Abstract

    e.  Table of contents

    f.  List(s) of figures and tables (if included)

    g.  Foreword (optional)

    h.  Preface (administrative information)

    i.  Acknowledgments (optional).

ANSI/NISO Z39.18-2005[2] states that a cover, a report documentation section (page), a foreword, a preface, and acknowledgments are optional or conditional; however, defense-related technical reports should include a cover (front and back), a report documentation page, and a preface, which includes administrative information.

## 3.1   Front Cover and Title Page

The front cover and title page are nearly identical and are, therefore, addressed together. The purpose of the cover is twofold: to provide basic information regarding the technical report and to protect the contents of the technical report. (Information on the front cover of a print technical report may become illegible over time because of exposure and repeated handling; thus, essential information on the front cover is repeated on the title page.) The title page repeats basically the same information as the front cover but may be more elaborative; therefore, the title page should be used as a reference source in lieu of the front cover. (The only significant difference between the front cover and title page occurs when an author or co-author is not an employee of the performing organization. This is addressed in Section 3.11.)

No page number should appear on the front cover or title page; however, the title page is considered page i. The back of the title page may be paginated in certain circumstances (Section 3.4).

The front cover and title page should contain the following items:

- Technical report number
- Document control number and serial number (if classified and applicable)
- Performing organization (including address)
- Publication date
- Title and subtitle (if subtitle is used)
- Title and numbering of series (if the technical report is issued in a series)
- Author(s)
- Notices
  - Overall (highest) classification level (if classified)
  - Unclassified controlled information (if unclassified and applicable)
  - Intelligence control markings (if classified and applicable)
  - Warning notice (if applicable)
  - Restrictive markings on noncommercial data and software (if applicable)
  - Distribution statement
  - Export control notice (if applicable)
  - Downgrade, declassify, and related markings (if classified)
  - Destruction notice
  - Foreign government and North Atlantic Treaty Organization (NATO) information (if classified and applicable).
- Seals/emblems and logos.

## 3.2    Inside Front Cover

The inside front cover may contain disclaimers or similar notices; otherwise, it should be blank. ANSI/NISO Z39.18-2005 provides the following advice regarding disclaimers or notices:

> If disclaimers or similar notices are needed, they appear on the inside front cover or the optional copyright page that follows the title page of a printed [technical] report. Notices may also alert the reader to certain legal conditions, for example, using brand or trade names in the [technical] report. Generic terms are preferable to brand or trade names if scientific and technical accuracy can be maintained in using them.

> A disclaimer may or may not be appropriate for government-generated [technical] reports. It is the responsibility of each organization to determine the appropriate notice for the [technical] reports it produces, and to coordinate these decisions with the appropriate legal counsel. Government classified material will have specific regulations; producers should follow the regulations applicable to their government agency.[2]

Because a front cover is required in a defense-related technical report, a disclaimer or similar notice, if used, should appear on the inside front cover.

For DoD contractor technical reports, the following statement or a comparably worded statement should, if required by the sponsoring Government organization, appear on the inside front cover of a technical report prepared under Government sponsorship:

> This report was prepared as an account of work sponsored by an agency of the United States Government. Neither the United States Government, nor any agency thereof, nor any of their employees, makes any warranty, express or implied, or assumes any legal liability or responsibility for the accuracy, completeness, or usefulness of any information, apparatus, product, or process disclosed, or represents that its use would not infringe privately owned rights. Reference herein to any specific commercial product, process, or service by trade name, trademark, manufacturer, or otherwise does not necessarily constitute or imply its endorsement, recommendation, or favoring by the United States Government or any agency thereof. The views and opinions of authors expressed herein do not necessarily state or reflect those of the United States Government or any agency thereof.[4]

For classified technical reports, **UNCLASSIFIED**[e] should appear at the top and bottom of the inside front cover, assuming that the information on the inside front cover is unclassified; if not, the highest classification of the information on the inside front cover should appear at the top and bottom. This also applies to the back of the title page. Notwithstanding, some organizations prefer or require that the overall classification of the technical report be placed on all pages.

## 3.3    Back Cover

ANSI/NISO Z39.18-2005[2] states that the technical report number and title and numbering of the series may appear on the back cover; however, this is unusual in defense-related technical reports and, therefore, optional. The back cover is usually blank on unclassified technical reports unless the technical report contains unclassified controlled information. (Unclassified controlled information is addressed in Section 3.12.2.) On classified technical reports, the back cover should contain the overall classification level of the technical report. The inside back cover should be blank on unclassified and classified technical reports. No page number should appear on the back cover or on the inside back cover.

## 3.4    Back of Title Page

The back of the title page should be blank on unclassified and classified technical reports unless the technical report is copyrighted. If copyright information appears on the back of the title page, the page should be numbered ii in the footer. Copyright information should not appear in the table of contents.

Under U.S. copyright law, copyright protection automatically subsists from the time an original work (technical report) is created in fixed form. The copyright immediately becomes the property of the author who created the technical report. No notice or formal registration is required.

U.S. copyright law excludes Government works prepared solely by officers and employees of the Government as part of their official duties from copyright protection. However, a privately created work, e.g., quotation, photograph, drawing, etc., included in a Government work does not place the private work in the public domain. ANSI/NISO Z39.18-2005 adds the following: "While federal government publications may be freely copied by

---

[e]Boldface type is used herein to distinguish markings and certain notices from the rest of the text.

the public, a request for permission allows the publisher to track the uses of the [technical] report."[2]

DoD contractors and grantees are not considered Government employees, even though they may be working under a Government contract.

> [T]he Contractor may establish, without prior approval of the Contracting Officer, claim to copyright subsisting in scientific and technical articles based on or containing data first produced in the performance of this contract and published in academic, technical or professional journals, symposia proceedings or similar works. . . . For data other than computer software the Contractor grants to the Government, and others acting on its behalf, a paid-up, nonexclusive, irrevocable worldwide license in such copyrighted data to reproduce, prepare derivative works, distribute copies to the public, and perform publicly and display publicly, by or on behalf of the Government.[21]

Further guidance regarding copyrights is provided in the Defense Federal Acquisition Regulation Supplement (DFARS).[22] A typical copyright notice would appear as follows:

> Copyright © 2006 [name of DoD contractor]. All rights reserved.

> This material may be used, modified, or reproduced by or for the U.S. Government pursuant to the license rights granted under the clauses at DFARS 252.227-7013/7014.

> For any other permissions, please contact the [appropriate office at DoD contractor].[f]

## 3.5    Technical Report Number

The technical report number should be different for each technical report and should comply with ANSI/NISO Z39.23-1997.

> The Standard Technical Report Number (STRN) shall be used with all technical reports, including those produced in nonprint media. The [technical] report number shall appear in an upper corner on both the [front] cover and title page and on the spine of a bound [technical] report if space permits so that a user will not have to remove the [technical] report from a shelf to read the number. A [technical] report

---

[f]The sample copyright notice is based on the copyright notice used by the Johns Hopkins University Applied Physics Laboratory (APL). The sample notice was provided by APL's Office of Patent Counsel.

number is composed of an alphanumeric [technical] report code (2–16 characters), a 2-character group separator, and sequential group of 1–16 characters indicating the year, sequence of [technical] report issuance, and identifying characters for supplements, revisions, drafts, etc., as appropriate. The [technical] report number shall appear on all copies of the [technical] report.[11]

As stated previously, the technical report number may also appear on the back cover. ANSI/NISO Z39.23–1997 identifies the "2-character group separator"[11] as a double hyphen; however, a single em dash may be used in lieu of a double hyphen.

The performing and sponsoring organizations may have separate technical report numbers; if so, both should be included on the front cover, title page, and back cover (optional) and an explanation should appear in the preface (administrative information). Section 3.7 provides additional information.

The following example is derived from the actual guidance on creating a technical report number for a DoN organization.[g]

### NNNNNN-CC-TR—YYYY/XX+CR

The technical report code comprises the following items:

**NNNNNN**—This represents the acronym of the performing/sponsoring organization. The short acronym of an organization should be used in lieu of the long acronym, e.g., NAVSEA, not NAVSEASYSCOM, for the Naval Sea Systems Command.

**CC**—This two-digit character identifies the performing/sponsoring sub-organization's code. The character limit may vary according to the organization.

**TR**—This is an acronym for "Technical Report." (This should be used if the technical report numbering format is also used for other similar documents, such as technical memorandums [**TM**]; otherwise, it may be omitted.)

An em dash separates the technical report code and sequential group.

The sequential group comprises the following items:

**YYYY**—This four-digit (not two-digit) numerical character identifies the calendar year (not fiscal year) the technical report was assigned a number, e.g., **2006**. Some organizations assign a technical report number at its inception; others wait until the final draft is complete.

[g]Enclosure (1), "Technical Report Numbering Guidance," to Memorandum from NSWCCD Commander/Director to Technical Directorate Heads, "Technical Report Numbers," 10 August 1998.

**XX**—This two-digit numerical character identifies the sequential number assigned to the technical report, e.g., **01, 02**, etc. If an organization expects to publish more than 99 technical reports in a calendar year, this should be a three-digit character, e.g., **001, 002**, etc.

**CR**—This is an acronym for "Contractor Report." These characters and the plus sign should be omitted if they are not applicable. In this instance, a contractor technical report is a technical report whose sole or primary author is working under a DoD contract or grant and the technical report is being published by DoD or a DoD component. DoD contractor technical reports published in-house do not need to include "CR" in their technical report numbers.

For example, the first DoD contractor-produced technical report originating in Code 63 of the Naval Surface Warfare Division, Carderock Division (NSWCCD) in 2006 would have the following technical report number:

**NSWCCD-63-TR—2006/01+CR**

Note that an en dash (or hyphen) serves as a subdivider within the technical report code. A slash separates the year and the sequential number within the sequential group. A plus sign separates the sequential number and "CR."

## 3.6    Document Control Number

Classified technical reports are required to be controlled and accounted for by the performing organization; therefore, the exact number of copies (paper and electronic) of the final technical report should be known by the performing organization. Many organizations assign an identification number or document control number to all externally distributed classified documents, including technical reports. The document control number is separate from the technical report number. The serial number accompanying the document control number indicates the number assigned that particular technical report; the exact number of copies of the technical report is indicated in the distribution list (Chapter 5, Section 5.6). Copies recorded on removable electronic media, e.g., a compact disk–read only memory (CD–ROM), should also include the document control number and be assigned a serial number. On paper technical reports, the document control number should appear on the front cover and title page, and the serial number should be typed or stamped underneath the document control number. On removable electronic media, the same information should appear on the label.

## 3.7    Performing Organization and Sponsoring Organization

The name and address (city, State, and ZIP code) of the performing organization should appear on the front cover and title page, usually at the top.
A suborganization of the performing organization may also appear on the
front cover and title page. Because of space limitations, the name and address of the sponsoring organization, if different from the performing organization, should appear only in the preface (administrative information)
(Section 3.20).

## 3.8    Publication Date

The publication date is the month and year (not day, month, and year) the
technical report is published (not necessarily assigned a technical report
number). Technical reports are sometimes subject to a lengthy review and
approval process; the final review constitutes the completion of the technical report. The technical report number should generally correspond to a
calendar year; therefore, if the period between the assignment of the technical report number and the publication date extends significantly beyond the
calendar year in which the technical report number was originally assigned,
the technical report number should be assigned at the completion of the
review process or a new technical report number should be assigned to correspond with the publication date. By limiting the publication date to month
and year only, as opposed to day, month, and year, the publication date is
more inclusive and, thus, requires fewer updates. Additionally, the publication date usually does not coincide with the date the related RDT&E was
completed, which may have preceded the publication date by months.

## 3.9    Title and Subtitle

The title is, in many ways, the most significant information in a technical
report. It is the first item a prospective reader sees, and if it is inadequate,
it may be the last. The title should be concise, without being too short and
ambiguous, and specific, without being too long and redundant. Subtitles
are used for clarity to distinguish a series of related technical reports and are
also an effective means to divide otherwise long titles.

Titles and subtitles should not contain unnecessary introductory phrases,
such as "Report on,"[2] "Investigation of," etc. Titles should not begin with
the indefinite articles "a" or "an" or the definite article "the" unless comprehension is affected; however, articles are effective transitional devices between titles and subtitles, e.g., "Environmental Life Cycle Management: *An*
Overview of U.S. Navy and Other Department of Defense Initiatives"[23] and

"DD 21A—*A* Capable, Affordable, Modular 21st Century Destroyer."[24] Also, titles and subtitles should not begin or end with terms such as "interim report" or "final report"; this information is indicated on the report documentation page (Standard Form 298) (Section 3.15) and in the preface (administrative information) (Section 3.20).

Abbreviations—specifically acronyms—should, ideally, not appear in titles and subtitles; instead, they should be spelled out to facilitate indexing and referencing. However, if an acronym is more commonly known than its spelled-out equivalent, the acronym may appear in parentheses after its spelled-out equivalent.

Classified technical reports should, if possible, have unclassified titles and subtitles so they may be referenced in unclassified, limited distribution publications. If a subtitle is used in a classified technical report, it may be marked separately from the title; however, in many instances, the subtitle is separated from the title by only a colon or a dash and is, thus, considered part of the title for classification purposes. If the subtitle is separate and distinct from the title, e.g., placed beneath the title, it should be marked separately. Regardless, the title and subtitle should, ideally, have the same classification level. The classification mark should appear in parentheses after the title and, in the latter case, the subtitle as well.

### 3.10    Titling and Numbering of Series

Occasionally, a technical report is one of a series of technical reports dealing with the same subject. If a broader title is used for the series, this title is added to the front cover and title page; if the broader title is the same as the title of the technical report, it is unnecessary to repeat it. The series number should indicate the number or volume of the technical report in the series and the total number of technical reports in the series, e.g., "Technical Report 2 of 5" or "Volume I of III."

### 3.11    Authorship

Defense-related technical reports usually fall into one of the following four scenarios pertaining to authorship:

a. A DoD or DoD component technical report authored by an employee(s) of the same or another DoD or DoD component organization or a combination of the two. Members of the Armed Forces are considered DoD employees. This scenario also includes non-DoD Government employees as co-authors.

b. A DoD or DoD component technical report authored by an employee(s) of the same or another DoD or DoD component organization or a combination of the two and co-authored by a DoD contractor employee(s). The DoD contractor employee(s) is not the principal author.

c.  A DoD or DoD component technical report authored by a DoD contractor employee(s) and co-authored by a DoD or DoD component organization employee(s). The DoD employee(s) is not the principal author.

d.  A DoD contractor technical report authored by a DoD contractor employee(s). DoD contractors from other DoD contractor organizations may co-author the technical report.

In scenarios a to c, the technical report is published as a DoD or DoD component organization technical report. Basically, if a DoD employee is an author or co-author, the technical report is usually published as a DoD or DoD component organization technical report. In addition, a DoD or DoD component organization technical report authored by a DoD contractor employee(s) (category c) may require co-authorship by a DoD employee.

The performing organization may be DoD or a DoD component organization or a DoD contractor organization; regardless, the distinction is between Government (DoD and DoD component organizations) and non-Government (academia and industry) authors and it affects copyright, as discussed in Section 3.4, and the inclusion of certain items on the front cover and title page of technical reports (Figures 3–1 to 3–9).

Defense-related technical reports also have a "corporate author," which is the performing/sponsoring organization discussed in Section 3.7 and a "personal author," who is discussed in this section.

ANSI/NISO Z39.18–2005 introduces the term "creator"[2] (of the technical report) to the standard. As used herein, "author" and "creator" are synonymous; only "author" will be used hereinafter. ANSI/NISO Z39.18–2005 states that the authorship of a technical report "is reserved for the person or persons responsible for originating the scientific or technical information or the text of the [technical] report and who can effectively defend the content of the [technical] report to a peer group. The primary author . . . is always identified first."[2] Co-authors should be listed in decreasing order of significance or input to the work; they should not be intentionally listed alphabetically. Other contributors to the technical report (not necessarily the RDT&E), such as editors and illustrators, should not be listed unless they have "applied subject matter expertise"[2] to the technical report. A maximum of 10 names may be listed.

The term "et al." (meaning "and others") should not be used in lieu of listing the co-authors. If the author or co-author is not an employee of the performing organization, the name and organization of the author or co-author should appear on the front cover and the name and address (city, State, and ZIP code) of the author's or co-author's organization should appear on the title page.

Names of authors should appear as follows: first name, middle name (if any) or initial (preferable), last name. "Mr.," "Mrs.," "Ms.," or "Dr." should not precede the name. Academic degrees are inappropriate; however, civilian job titles are acceptable[2] and, if included, should appear after or underneath the name.

Military titles are also acceptable to distinguish civilians from the military, and the spelled-out or abbreviated military title should precede the name. (Table 3-1 lists military title abbreviations for commissioned and warrant officers of the Army, Air Force, Marine Corps, Navy, and Coast Guard.) A complete list of military title abbreviations is in the *United States Government Printing Office Style Manual 2000*.[20]

**Table 3-1.  Military Title Abbreviations**

| Army, Air Force, and Marine Corps | | Navy and Coast Guard | |
|---|---|---|---|
| Title | Abbreviation | Title | Abbreviation |
| General | GEN | Admiral | ADM |
| Lieutenant General | LTG | Vice Admiral | VADM |
| Major General | MG | Rear Admiral | RADM* |
| Brigadier General | BG | Commodore† | COMO |
| Colonel | COL | Captain | CAPT |
| Lieutenant Colonel | LTC | Commander | CDR |
| Major | MAJ | Lieutenant Commander | LCDR |
| Captain | CPT | Lieutenant | LT |
| First Lieutenant | 1LT | Lieutenant Junior Grade | LTJG |

*"(UH)" and "(LH)" may be placed after "RADM" to distinguish upper half and lower half, respectively.
†Not currently in use.
‡No warrant officers in Air Force.

**Table 3-1.   Military Title Abbreviations (Continued)**

| Army, Air Force, and Marine Corps | | Navy and Coast Guard | |
|---|---|---|---|
| Title | Abbreviation | Title | Abbreviation |
| Second Lieutenant | 2LT | Ensign | ENS |
| Warrant Officer[‡] | WO | Warrant Officer | WO |

*"(UH)" and "(LH)" may be placed after "RADM" to distinguish upper half and lower half, respectively.
[†]Not currently in use.
[‡]No warrant officers in Air Force.

## 3.12    Notices[h]

### 3.12.1    Overall Classification Level

If the technical report is classified, the overall classification level, i.e., **TOP SECRET**, **SECRET**, or **CONFIDENTIAL** (all-capital letters), should appear at the top and bottom center of the front and back covers and title page in a type larger than any other on the front and back covers and title page so that the classification markings are conspicuous.

The size, type, and color of the markings do not necessarily make the markings conspicuous. A marking is conspicuous when it is noticed and recognized by the holder as separate or different from other information or material; it, thus, warns of the special requirements necessary for protection. No other terms should be used to identify a classified technical report. Unclassified technical reports should not be marked **UNCLASSIFIED** unless the technical report has been examined specifically for classification purposes.

### 3.12.2    Unclassified Controlled Information

There are types of unclassified information that require control beyond the normal requirements for unclassified information; they fall into the general category of unclassified controlled information. While it is unlikely that all

---

[h]Unless indicated otherwise, the information in Section 3.12 was obtained from the *STI Handbook*,[5] DOD 5200.1-PH,[25] and DOD 5200.1-R.[26]

types of unclassified controlled information will appear in a defense-related technical report, the following are described in DOD 5200.1-R[26] and presented here:

- For Official Use Only (FOUO) information.
- Sensitive But Unclassified (SBU) (formerly Limited Official Use) information. This originates in the Department of State.
- Drug Enforcement Administration (DEA) sensitive information. DEA is part of the Department of Justice.
- DoD Unclassified Controlled Nuclear Information (DoD UCNI).
- Sensitive information (Computer Security Act of 1987).

As prescribed herein, all unclassified controlled information markings, except for DEA sensitive information, should be placed at the bottom of the front cover, title page, back cover, and each page containing unclassified controlled information. Portion markings are required for FOUO information in unclassified technical reports. DEA sensitive information should be marked at the top and bottom of the same pages.[i]

Classified technical reports may contain unclassified controlled information, and those portions and pages should be indicated. (Chapter 4, Figures 4-6 and 4-8 [pages 165 and 173, respectively] show page and portion markings.) However, the overall classified technical report should not be considered unclassified controlled information, and, thus, these notices should not appear on the front cover, title page, and back cover of a classified technical report.

### 3.12.2.1  For Official Use Only (FOUO)

FOUO is the most common type of unclassified controlled information to appear in defense-related technical reports. FOUO is not a classification; rather, it indicates sensitive but unclassified information exempt from mandatory public release under the Freedom of Information Act (FOIA). Specifically, FOUO designates "information that has not been given a security classification pursuant to the criteria of an Executive Order, but which may be withheld from the public *because disclosure would cause a foreseeable harm to an interest protected by*"[27] exemptions to the FOIA. Information exempt under the Privacy Act should also be marked FOUO.

While classified technical reports are exempt from mandatory public release under the FOIA (Title 5 U.S. Code [U.S.C.] §552[b][1]), they should not be additionally marked FOUO as an overall classification: "No other

---

[i]Figures herein that refer to unclassified controlled information do not indicate this additional requirement for DEA sensitive information.

material shall be considered FOUO and FOUO is not authorized as an ane-
mic form of classification to protect national security interests."[27] However,
as previously stated, unclassified portions of a classified technical report
may be additionally assigned an FOUO designation.

The FOUO designation should only be applied to unclassified technical
reports if one or more of the following exemptions[28] to public release under
the FOIA are applicable:

- Information that pertains solely to the internal personnel rules and
  practices of DoD (Title 5 U.S.C. §552 [b][2]). This exemption is entirely discretionary.
- Information specifically exempted by statute or where the statute establishes particular criteria for withholding (Title 5 U.S.C. §552
  [b][3]). The language of the statute must clearly state the information
  will not be disclosed; therefore, there is no discretion.
- Information such as trade secrets or commercial or financial information obtained from a person or organization outside the Government
  on a privileged or confidential basis when disclosure of the information is likely to cause competitive harm to the company, impair the
  Government's ability to obtain necessary information in the future,
  or impair some other legitimate Government interest (Title 5 U.S.C.
  §552 [b][4]). There is no discretion.
- Interagency and intraagency memoranda or other internal documents
  that are deliberative in nature—containing subjective evaluations,
  opinions, internal advice, and recommendations—and are also part
  of the decisionmaking process (Title 5 U.S.C. §552 [b][5]). This exemption concerns documents that would normally be considered
  "privileged" or protected from release during the discovery process
  of a civil lawsuit when litigants obtain information from one another.
  An organization's legal counsel should be consulted to determine
  whether the record in question would normally be made available
  through discovery. This type of information is in contrast to factual
  matters that may be the basis on which an evaluation or advice is
  provided. Such facts may also be withheld if they are not severable
  or if protected by other exemptions. This exemption is entirely discretionary.
- Personal information contained in a Privacy Act system of records
  (including personnel and medical files) that, if disclosed to a request-
  or other than the person whom the information is about, would result
  in a clearly unwarranted invasion of personal privacy and could subject the releaser to civil and criminal penalties (Title 5 U.S.C. §552
  [b][6]). The information must be identifiable to a specific individual

and not freely available from sources other than the Federal Government (thus requiring FOIA action by third party requester). There is no discretion.

- Records or information compiled for law enforcement purposes, i.e., civil, criminal, or military law, including the implementation of Executive orders or regulations issued pursuant to law (Title 5 U.S.C. §552 [b][7]). This exemption is the law enforcement counterpart of the previous exemption and is discretionary with the exception of an unwarranted invasion of personal privacy of a living person or surviving family members of the person identified in the record or could reasonably be expected to endanger the life or physical safety of any individual.
- Certain records related to the examination, operation, or condition reports prepared by, on behalf of, or for use by an agency for regulation of supervision of financial institutions (Title 5 U.S.C. §552 [b][8]).
- Geological and geophysical information and data (including maps) concerning wells (Title 5 U.S.C. §552 [b][9]).

Examples of the preceding exemptions are provided in AD-A 423 966.[28]
As stated previously, FOUO technical reports, like other unclassified controlled information, should be marked in a manner similar to classified technical reports.

An unclassified [technical report] containing FOUO information shall be marked "For Official Use Only" at the bottom on the outside of the front cover (if any), on each page containing FOUO information [not necessarily every page], and on the outside of the back cover. *Each paragraph* [portion] *containing FOUO information shall be marked as such.*[j] [The title page should be marked like the front cover. If possible, a date or event when the marking may be removed should be included; Section 3.12.8 provides guidance.]

Within a classified [technical report], an individual page that contains both FOUO and classified information shall be marked at the top and bottom with the highest security classification of information appear-

---

[j]Chapter 4, Figure 4-1 (page 148), Figure 4-3 (page 152), Figure 4-5 (page 159), and Figure 4-7 (page 167) show examples of FOUO portion marking in the executive summary and main body of an unclassified technical report. Chapter 5, Figure 5-1 (page 190), Figure 5-3 (page 193), Figure 5-5 (page 197), and Figure 5-7 (page 200) show examples of FOUO portion markings in an unclassified appendix.

ing on the page. Individual paragraphs [portions] shall be marked at the appropriate level, as well as unclassified or FOUO, as appropriate.

Within a classified [technical report], an individual page that contains FOUO information but no classified information shall be marked "For Official Use Only" at the top and bottom of the page, *as well as each paragraph* [portion] *that contains FOUO information.*

FOUO material transmitted outside the Department of Defense requires application of an expanded marking to explain the significance of the FOUO marking. This may be accomplished by typing or stamping the following statement on the record prior to transfer:
**This document contains information exempt from mandatory disclosure under the FOIA.**
**Exemption(s) [insert applicable citation(s), e.g., 5 U.S.C. 552 (b)(3)] applies/apply.**

FOUO information may be disseminated within DoD Components and between officials of DoD Components and DoD contractors, consultants, and grantees to conduct official business for the Department of Defense. Recipients shall be made aware of the status of such information, and transmission shall be by means that preclude unauthorized public disclosure. Transmittal documents shall call attention to the presence of FOUO attachments.[27]

### 3.12.2.2    Sensitive But Unclassified (SBU) Information

SBU information originates in the Department of State and is equivalent to FOUO information and should be marked equivalently in DoD technical reports.

### 3.12.2.3    Drug Enforcement Administration (DEA) Sensitive Information

DEA sensitive information originates in DEA. DoD unclassified technical reports containing this information should be marked **DEA Sensitive** at the top and bottom of the front cover, title page, and back cover. Each page of an unclassified DoD technical report containing DEA sensitive information should also be marked **DEA Sensitive** at the top and bottom of the page; classified DoD technical reports should be marked the same if the individual page is unclassified.

### 3.12.2.4    DoD Unclassified Controlled Nuclear Information (DoD UCNI)

DoD UCNI is unclassified information on security measures (including security plans, procedures, and equipment) for the physical protection of DoD Special Nuclear Material (SNM), equipment, or facilities. Information is designated DoD UCNI only when it is determined that its authorized disclosure could reasonably be expected to have a significant adverse effect on the health and safety of the public or the common defense and security by increasing significantly the likelihood of the illegal production of nuclear weapons or the theft, diversion, or sabotage of DoD SNM, equipment, or facilities.

The front cover, title page, and back cover of unclassified technical reports should be marked **DoD Unclassified Controlled Nuclear Information** if they contain DoD UCNI. Any other page containing DoD UCNI information should also be marked. DoD UCNI material transmitted outside DoD should contain the following expanded marking on the front cover and title page:

**Department of Defense**
**Unclassified Controlled Nuclear Information**
**Exempt from Mandatory Disclosure**
**(5 U.S.C. 552 [b][3], as authorized by 10 U.S.C. 128)**

### 3.12.2.5    Sensitive Information (Computer Security Act of 1987)

The Computer Security Act of 1987 established requirements for the protection of certain information in the Federal Government's automated information systems. This information is referred to as "sensitive" and is defined in the Act. "Sensitive" information guidelines are similar to FOUO guidelines.

The front cover, title page, and back cover of unclassified technical reports should be marked **Sensitive** if they contain this type of information.

### 3.12.3    Intelligence Control Markings

In addition to the overall classification level, the following intelligence control markings are included, if applicable, on the front cover and title page of classified technical reports.

- **Authorized for Release to (name of country or international organization)**. This marking should be used on classified information that an originator has determined can be released to the indicated foreign countries or organizations through established foreign disclosure procedures and channels. The name of the country or international organization should not be abbreviated on the front cover and title page. DODD 5230.11[29] provides additional guidance regarding the disclosure of classified military information to foreign governments and international organizations.

- **Caution—Proprietary Information Involved**. This marking identifies classified information provided by a commercial firm or private source with the understanding that the information will be protected as a trade secret or proprietary data.

- **Dissemination and Extraction of Information Controlled by Originator**. This marking is used so the originator can continually supervise use of the information. This marking should be used only on classified intelligence information that clearly identifies or would permit ready identification of an intelligence source or method that is susceptible to countermeasures that could nullify or measurably reduce its effectiveness. This marking should not be used when the information can be reasonably protected by using other markings or by using the "need to know" principle of the security classification system.

- **Not Releasable to Contractors/Consultants**. This marking prohibits the release of classified technical reports to contractors and consultants without the permission of the originator. This marking should be used when release of the information to a contractor or consultant would provide a competitive advantage that could conflict with the contractor's or consultant's obligation to maintain the security of the information. This notice should also be used on material provided by a source, with the condition that it not be made available to contractors or consultants.

- **Not Releasable to Foreign Nationals**. This marking identifies classified intelligence information that may not be released in any form to foreign governments, foreign nationals, or non-U.S. citizens without permission of the originator. This marking should be used on intelligence information that could jeopardize intelligence sources or methods if released to a foreign government or national. This marking should also be used on information that should not be released because of U.S. policy.

## 3.12.4    Warning Notices

A warning notice or an export control notice (Section 3.12.7) may be included on the front cover and title page and advises users that a classified technical report (or an unclassified technical report containing naval nuclear propulsion information [NNPI]) requires additional protective measures in addition to those imposed by the classification level. The classification source document usually provides information about which warning notice, if any, must be used. The following warning notices may appear on technical reports:

- Communications Security (COMSEC) material—If the classified technical report contains COMSEC material and it is being released to contractors, the following warning notice should be included on the front cover and title page:
  **COMSEC Material—Access by contractor personnel restricted to U.S. citizens holding final Government clearance.**

- Critical Nuclear Weapon Design Information (CNWDI)—If the classified technical report contains CNWDI, the following warning notice should be included on the front cover and title page:
  **Critical Nuclear Weapon Design Information—DoD Directive 5210.2[30] applies.**

- Distribution and duplication notice—If the classified technical report is subject to special distribution and duplication limitations, one of the following two warning notices should be included on the front cover and title page:
  **Reproduction requires approval of originator or higher DoD authority.**
  or
  **Further dissemination only as directed by (insert appropriate command or official) or higher DoD authority.**

- NNPI—If the classified or unclassified technical report contains NNPI, one of the following two warning notices should be included on the front cover and title page:
  **Special Handling Required—Not releasable to foreign nationals.**
  or
  **This document (or material) is subject to special export controls, and each transmittal to foreign governments or foreign nationals may be made only with prior approval of (originating command).**

- Restricted data and formerly restricted data (atomic energy information)—If the classified technical report contains restricted data or formerly restricted data pertaining to atomic energy, one of the following two warning notices should be included on the front cover and title page:
  **RESTRICTED DATA—This document contains RESTRICTED**

DATA as defined in the Atomic Energy Act of 1954. Unauthorized disclosure subject to administrative and criminal sanctions.

or

FORMERLY RESTRICTED DATA—Unauthorized disclosure
subject to administrative and criminal sanctions. Handle as RE-
STRICTED DATA in foreign dissemination; Section 144b,
Atomic Energy Act of 1954.

### 3.12.5 Restrictive Markings on Noncommercial Data and Software

Noncommercial data and software affect Government data rights (licenses)
and are a factor in determining the appropriate distribution statement for a
technical report. If applicable, restrictive markings should appear on the
front cover and title page of classified and unclassified technical reports.

Restrictive markings are required on all noncommercial technical data
and computer software being delivered with less-than-unlimited
rights. The DFARS establishes specific procedures governing the
placement of restrictive markings on deliverables, storage media, and
transmittal documents. In addition, there are only six types of legends
that are authorized under the clauses:

- A notice of copyright under 17 U.S.C. 401 or 402,
- The GPR [Government Purpose Rights] legend,
- The limited-rights legend,
- The restricted-rights legend,
- The special-license-rights legend, and
- Pre-existing markings authorized under a previous Government con-
  tract.

The DFARS clauses specify the precise wording of the legends. Any
alterations of the prescribed content or format result in the marking
being considered "nonconforming."[31]

DFARS Subpart 227.71[32] governs the license rights and restrictive markings for technical data, and DFARS Subpart 227.72[33] governs computer
software and computer software documentation. The specific legends
(markings) are identified in the applicable clauses:

- DFARS 252.227-7013[34] for technical data—noncommercial items
- DFARS 252.227-7014[35] for noncommercial computer software and
  noncommercial computer software documentation
- DFARS 252.227-7018[36] for noncommercial technical data and computer software—Small Business Innovation Research (SBIR) Program.

## 3.12.6    Distribution Statements[k]

Distributions statements should appear on the front cover and title page of all unclassified and classified technical reports. The purpose of a distribution statement is to control the technical report's secondary distribution by indicating who can obtain the technical report without approval by the controlling DoD office. Primary distribution is indicated by the distribution list, which is discussed in Chapter 5, Section 5.6.

Distribution statements comprise four components: audience, reason(s) for limitation, date of determination, and controlling DoD office. The controlling DoD office is the organization that sponsored the RDT&E that generated the technical report or received the technical report on behalf of DoD and, therefore, has responsibility for determining the distribution availability of the technical report. (The specific code within the controlling DoD office is usually included in parentheses in the distribution statement following the name of the controlling DoD office.)

Authors are responsible for making an initial determination regarding the assignment of proper distribution statements to their technical reports. DTIC will not accept a technical report without a distribution statement. The decision to assign a specific distribution statement is based on the work, itself, as it applies to criteria for the distribution statements; the author's knowledge of the sponsoring organization's requirements; and the overall security requirements of the work. Table 3-2 lists the seven distribution statements.

### Table 3-2.    Distribution Statements

| | |
|---|---|
| Distribution Statement A: | Approved for public release; distribution is unlimited. |
| Distribution Statement B: | Distribution authorized to U.S. Government agencies only; (fill in reason); (insert date of determination). Other requests for this document shall be referred to (insert controlling DoD office). |
| Distribution Statement C: | Distribution authorized to U.S. Government agencies and their contractors; (fill in reason); (insert date of determination). Other requests for this document shall be referred to (insert controlling DoD office). |

---

[k]In addition to the *STI Handbook*, the information in Section 3.12.6 was obtained from DODD 5230.24[37] and AD-A 423 966.[28]

**Table 3-2.    Distribution Statements (Continued)**

Distribution Statement D:    Distribution authorized to the Department of Defense (DoD) and U.S. DoD contractors only; (fill in reason); (insert date of determination). Other requests for this document shall be referred to (insert controlling DoD office).

Distribution Statement E:    Distribution authorized to Department of Defense components only; (insert reason); (insert date of determination). Other requests for this document shall be referred to (insert controlling DoD office).

Distribution Statement F:    Further dissemination only as directed by (insert controlling DoD office) or higher DoD authority; (insert date of determination).

Distribution Statement X:    Distribution authorized to U.S. Government agencies and private individuals or enterprises eligible to obtain export-controlled technical data in accordance with (insert appropriate regulation); (insert date of determination). Controlling DoD office is (insert controlling DoD office).

## 3.12.6.1    Distribution Statement Checklist

Table 3-3 is a checklist indicating the reasons for limiting distribution of a technical report and which distribution statements they apply to. Only those reasons for limiting distribution listed in Table 3-3 and defined in Section 3.12.6.2 should be used; new, different reasons should not be "created." Generally, not more than two reasons should be used in the same distribution statement. While some reasons, e.g., "foreign government information" and "critical "technology," may be used in all of these distribution statements, other reasons are limited to certain distribution statements. For example, "test and evaluation" may be used for distribution statements B and E; however, it should not be used for distribution statements C and D.

**Table 3-3.    Distribution Statement Checklist**

| Reasons for Limiting Distribution | Distribution Statements | | | | | | |
|---|---|---|---|---|---|---|---|
| | A | B | C | D | E | F | X |
| Public and Foreign Release* | ✔ | | | | | | |
| Foreign Government Information | | | ✔ | ✔ | ✔ | ✔ | |

*Unlimited distribution.

Table 3-3.    Distribution Statement Checklist (Continued)

| Reasons for Limiting Distribution | Distribution Statements | | | | | | |
|---|:---:|:---:|:---:|:---:|:---:|:---:|:---:|
| | A | B | C | D | E | F | X |
| Proprietary Information | | ✓ | | | ✓ | | |
| Critical Technology | | ✓ | ✓ | ✓ | ✓ | | ✓ |
| Test and Evaluation | | ✓ | | | ✓ | | |
| Contractor Performance Evaluation | | ✓ | | | ✓ | | |
| Premature Dissemination | | ✓ | | | ✓ | | |
| Administrative/Operational Use | | ✓ | ✓ | ✓ | ✓ | | |
| Software Documentation | | ✓ | ✓ | ✓ | ✓ | | |
| Specific Authority (Identify) | | ✓ | ✓ | ✓ | ✓ | | |
| Direct Military Support | | | | | ✓ | | ✓ |
| Paragraph 4-505 of DOD 5200.1-R[26] or Other Specific Authority | | | | | | ✓ | |
| Export Control of Unclassified Documents (Private Individuals or Enterprises) | | | | | | | ✓ |

*Unlimited distribution.

## 3.12.6.2    Definitions of Reasons for Limiting Distribution of Technical Reports

The reasons for limiting distribution of technical reports (distribution statements B, C, D, E, and X) are defined as follows:

- Foreign government information—This reason protects and limits distribution in accordance with the desires of the foreign government that furnished the technical information. Information of this type is normally classified at the Confidential or higher level.
- Proprietary information—This reason protects information not owned by the U.S. Government and is protected by a contractor's limited rights statement or is received with the understanding that it not be routinely transmitted outside the Government.
- Critical technology—This reason protects information and technical data that advance current technology, describe new technology in an area of significant or potentially significant military application, or relate to a specific military deficiency of a potential adversary. Information of this type may be classified or unclassified; when unclassified, it is controlled by the export laws and is subject to the provi-

sions of DoD Directive 5230.25.[38] An export control notice (Section 3.12.7) should accompany a distribution statement containing this reason on an unclassified technical report.

- Test and evaluation—This reason protects the results of test and evaluation of commercial products or military hardware when such disclosure may cause unfair advantage or disadvantage to the manufacturer of the product. An export control notice (Section 3.12.7) should accompany a distribution statement containing this reason.
- Contractor performance evaluation—This reason protects information in management reviews, records of contract performance evaluation, or other advisory documents that evaluate programs of contractors.
- Premature dissemination—This reason protects patentable information from premature dissemination on systems or processes in the development or conceptual stage.
- Administrative/operational use—This reason protects technical or operational data or information from automatic dissemination under the International Exchange Program or by other means. This protection covers technical reports required solely for official use or strictly for administrative or operational purposes. This statement can be applied to technical reports containing valuable technical or operational data.
- Software documentation—This reason protects software documentation that is to be released only in accordance with the software license.
- Specific authority—This reason protects information not specifically included in the other reasons but which requires protection in accordance with valid documented authority, such as Executive orders, classification guidelines, or DoD or DoD component regulatory documents. When filling the reason, the "specific authority (identification of valid documented authority)" should be cited.
- Direct military support—This reason protects technical reports containing export-controlled technical data of such military significance that release for purposes other than direct support of DoD-approved activities may jeopardize an important technological or operational military advantage of the United States. Designation of such data is made by a competent authority in accordance with DoD Directive 5230.25.[38] An export control notice (Section 3.12.7) should accompany a distribution statement containing this reason.

### 3.12.6.3    Date of Determination

The date of determination is the date the distribution statement is assigned,[28] which usually coincides with the publication date (Section 3.8). However, it may be a preceding date linked to certain information in the technical report, such as the date of a contract or other agreement. However, the date of determination should never be a date subsequent to the publication date, such as the date of submission of the technical report to the controlling office or DTIC.

### 3.12.6.4    Further Requests to Controlling Office

The full name and address of the controlling office should be provided, not just the controlling office's acronym. The controlling office is the organization that sponsored the work that resulted in a technical report. The controlling office may be a DoD or other Government organization or a foreign government or international alliance, such as NATO. The controlling office is never a contractor. The COTR should direct a contractor as to the appropriate DoD distribution statement for a technical report.

### 3.12.6.5    Use of Distribution Statements

### 3.12.6.5.1    Distribution Statement A

Distribution statement A should be used only on unclassified technical reports that have been cleared for public release by the performing or sponsoring organization. (DODD 5230.9[3] addresses the clearance of DoD information for public release.) Distribution statement A should not be used on classified technical reports, and it should not be used on technical reports that were formerly classified unless these technical reports have been cleared for public release. Also, distribution statement A should not be used on technical reports that contain classified references or classified bibliographic entries, technical reports that contain unclassified controlled information, or technical reports that contain export-controlled technical data.

Technical reports resulting from contracted fundamental or basic research efforts are usually assigned distribution statement A; exceptions occur when there is a high probability of disclosing performance characteristics of military systems or of manufacturing technologies that are unique and critical to national defense. These exceptions should be recorded in the contract or grant. Technical reports with this statement can be exported and are sold to the public, foreign nationals, foreign companies, and foreign governments.

Distribution statement A does not require a date, identification of the controlling office, or reason(s) for restriction.

### 3.12.6.5.2 Distribution Statements B, C, D, and E

Distribution statements B, C, D, and E may be used on classified and unclassified technical reports. These distribution statements are used on classified technical reports to provide distribution limitations in addition to "need to know" requirements and to serve as a controlling statement if the technical report is declassified. Distribution statement E should be used on technical reports that contain export-controlled information that is part of direct military support; an export control notice (Section 3.12.7) should accompany distribution statement E when it contains this reason.

### 3.12.6.5.3 Distribution Statement F

Distribution statement F is the most restrictive of the distribution statements, allowing the technical report to only be released by the DoD controlling office on a case-by-case basis. Distribution statement F is normally used only on classified technical reports; however, it may be used on unclassified technical reports when specific authority exists. This statement is used when the DoD originator determines that the information is subject to special dissemination limitations specified by paragraph 4-505 of DOD 5200.1-R.[26] When a classified technical report assigned distribution statement F is declassified, the statement is retained until the controlling DoD office assigns another distribution statement. No audience or reason(s) for restriction is specified in distribution statement F.

### 3.12.6.5.4 Distribution Statement X

Distribution statement X should be used on unclassified technical reports when distribution statements B, C, D, E, or F are not applicable and the technical report contains export-controlled technical data, as explained in DoD Directive 5230.25.[38] This statement should not be used on classified technical reports; however, it may be assigned to technical reports that were formerly classified. The only reasons for using distribution statement X are "critical technology" or "direct military support"; however, these reasons should not be inserted in the distribution statement. Instead, the appropriate regulation, e.g., legal statute or congressional act, should be cited; DoD Directive 5230.25 may also be cited. An export control notice (Section 3.12.7) should accompany distribution statement X.

### 3.12.7    Export Control Notice

An export control notice should be included on the front cover and title page of classified and unclassified technical reports if the technical report meets all of the following criteria:

- It is owned or controlled by DoD. (This includes technical reports prepared under DoD contracts or other agreements.)
- It has a military or space application, i.e., the information can be used (or adapted for use) to design, engineer, produce, manufacture, operate, repair, overhaul, or reproduce military or space equipment or related technology.
- It may be exported only with approval, authorization, or license under U.S. export control laws. (This applies to technical information related to items on the Department of State's U.S. Munitions List and the Department of Commerce's Commerce Control List.)
- It discloses critical technology, i.e., it reveals production "know-how" that would significantly contribute to a foreign country's military potential and possibly prove detrimental to the security of the United States. Such information usually occurs in the following areas:
  - Arrays of design and manufacturing know-how. This includes the know-how and related technical information required to achieve a significant development, production, or utilization purpose. Included are services, processes, procedures, specifications, design data and criteria, and testing techniques.
  - Keystone equipment. This includes manufacturing, inspecting, or testing equipment specifically necessary for the effective application of a significant array of technical information and know-how.
  - Keystone materials. These include materials specifically necessary for the effective application of a significant array of technical information and know-how.
  - Products accompanied by sophisticated know-how. Use of the products requires providing (disclosing) a significant amount of technical information and know-how (including operation, application, or maintenance know-how). Also included are products for which the embedded know-how can be derived by reverse engineering or is revealed by use of the products.

Authors should use a number of sources to determine if their technical reports require export controls. The two primary sources are the Department of State's U.S. Munitions List and the Department of Commerce's Commerce Control List. Secondary sources include the following:

- Developing Science and Technologies List
- Arms Export Control Act
- Export Administration Act of 1979
- International Traffic in Arms Regulations
- Export Administration Regulations
- Information from the intelligence and security communities
- Work in academia and industry
- Technology developments by U.S. Allies.

The Militarily Critical Technologies List (MCTL) is not a control list and should not be used to make an export control determination. However, the MCTL may be used as a reference source to assist in the process. The MCTL provides a detailed discussion of the development, production, and utilization of technologies that DoD has determined to be crucial to U.S. military capability and of significant value to potential adversaries. The MCTL is derived from many sources and includes the following technologies: technology not already possessed by a potential adversary; technology providing advantages to the United States in terms of performance, reliability, maintenance, and cost over systems currently in use by an adversary; and technology related to emerging technologies with high potential for having an impact on advanced military applications.

Critics of the MCTL have stated that it is "too broad, difficult to use, and out of date"[39]; therefore, it should not be relied upon as the sole source of information in this regard. DTIC posts the latest unclassified portions of the MCTL at the following Web site: www.dtic.mil/mctl.

If export control is required, the following notice should be used:

**WARNING—This document contains technical data whose export is restricted by the Arms Export Control Act (Title 22 U.S.C., Sec. 2751 et seq.) or the Export Administration Act of 1979, as amended (Title 50 U.S.C., App. 2401 et seq.). Violations of these export laws are subject to severe criminal penalties. Disseminate in accordance with the provisions of DoD Directive 5230.25.**

In lieu of referencing DOD 5230.25, some organizations reference an equivalent DoD component directive/instruction, e.g., Office of the Chief of Naval Operations Instruction (OPNAVINST) 5510.161.[40]

If an export control notice is included in a technical report, a separate notice (Enclosure 5 to DoD Directive 5230.25[38]) should appear in the preface (administrative information) (Section 3.20).

## 3.12.8    Downgrade, Declassify, and Related Markings[l]

The following information should be included on the front cover and title page of classified technical reports:

- Classification source—This is the source used to determine the technical report's overall classification level. It may be an individual, a source document, or a security classification guide.
- Reason for classifying—Each original classification decision should state a concise reason for classifying. The appropriate reasons (classification categories) are listed in Section 1.4 of Executive Order 13292.
- Date of origin—This is the publication date (Section 3.8) of the technical report.
- Office of origin—This is the originating organization. (The originating organization is the same as the performing organization [Section 3.7].)
- Downgrade information—This is the date or event when the overall classification level of the technical report will be changed to a lower classification level. (This is not always used; it depends upon the information provided in the source document.)
- Declassification date or event—This is the date or event when the technical report will be reviewed for declassification or the date or event when the technical report will be declassified.

Downgrade, declassify, and related markings should appear as follows on the front cover and title page:

- **Classified by:**_____(for originally classified material)

  or

  **Derived from:**_____(for derivatively classified material)

  These markings identify the classification source for the technical report. The classification source may be the original classification authority if the information in the technical report is originally classified, or it may be the source document, security classification guide (title or contract number, issuing agency, and date included), or Contract Security Classification Specification (for DoD contractors) if

---

[l]In addition to the *STI Handbook*[5] and DOD 5200.1-PH,[25] the information in Section 3.12.8 was obtained from DOD 5220.22-M,[41] Executive Order 13292,[18] and the Information Security Oversight Office's "Marking Classified National Security Information."[42] Quotations are specifically cited.

the technical report is derivatively classified. Most DoD classified material is derivatively classified; therefore, the latter marking will probably be used. If the technical report is originally classified, the **Classified by:** line should include the name or personal identifier of the actual classifier and his/her position. The date of the source document or security classification guide should also be included, if applicable. If more than one source document or security classification guide is used, **Multiple Sources** should be inserted. The multiple sources (source documents or security classification guides) should be identified and listed in bibliographic form in the preface (administrative information) (Section 3.20).

- **Reason classified:**_____(for originally classified material)

  The reason may be briefly described, e.g., **military plans; foreign relations** or equivalently cited from Executive Order 13292,[18] e.g., [Section] **1.4 (a) and (d)**. (A complete list of reasons from Executive Order 13292 is provided in Chapter 1, Section 1.7.1, page 21.)

- **Downgrade to:**_____ **on** _____

  This marking is not always used; it depends upon the instructions in the source document, security classification guide, or Contract Security Classification Specification (for DoD contractors). **SECRET** or **CONFIDENTIAL** (all-capital letters), as appropriate, should be inserted after **Downgrade to:**, and an effective date (day, month, year) or event should be inserted after **on**. Technical reports containing Restricted Data or Formerly Restricted Data (atomic energy information) should not contain this marking.

- **Declassify on:**_____

  The purpose of this marking is to provide any declassification instructions appropriate for the technical report. The date (day, month, and year) or event the technical report will be reviewed for declassification or the date or event when the technical report will be declassified is inserted. "If the original classification authority cannot determine an earlier specific date or event for declassification, information shall be marked for declassification 10 years from the date of the original decision, unless the original classification authority otherwise determines that the sensitivity of the information requires that it shall be marked for declassification for up to 25 years from the date of the original decision."[18]

  Notwithstanding, Executive Order 13292 allows for exemptions beyond the maximum 25-year classification period: "An original classification authority may extend the duration of classification, change the level of classification, or reclassify specific information only when the standards and procedures for classifying information under this order are followed."[18] Exemptions beyond the maximum 25-year classification period are granted only by the Interagency Security Classification Appeals Panel (ISCAP), which was created by Executive Order 12958.[43] When used, the **Declassify on:** line would include **25X** followed by the exemption category number from Executive Order 13292 and a date or event for declassification, e.g., **25X1, 15 May 2046**, which pertains to the first exemption category and "reveal[s] the identity of a confidential human source, or a human intelligence source, or reveal[s] information about the application of an intelligence source or method."[18]

Executive Order 12958, which was amended by Executive Order 13292, and DOD 5200.1-PH address eight exemption categories to the 10-year declassification rule. These exemption categories have been amended by Executive Order 13292, which has nine exemption categories to the 10-year declassification rule. The exemption categories are not parallel between the two Executive orders. For example, the first exemption category in Executive Order 12958 addresses "cryptologic systems,"[43] while the same subject is addressed by the third exemption category in Executive Order 13292. Notwithstanding, the exemption categories from Executive Order 12958 remain relevant to technical reports that are now derivatively classified but contain information that was originally classified and used one or more of the exemption categories from Executive Order 12958. If a derivatively classified technical report contains one or more of the exemption categories from Executive Order 12958, the author should identify the classification source document, indicate that the classification source document was marked with one or more exemption categories from Executive Order 12958, and include the date of the source document, e.g., **Derived from: Name of document (or classification guide used); Declassify on: Source marked X3; Date of source, 15 March 2003**.

Originally classified technical reports prepared after 22 September 2003 should use one or more of the exemption categories in Executive Order 13292, if applicable; they should not use the exemption categories listed in Executive Order 12958 or DOD 5200.1-PH, i.e., "X1" to "X8." The exemption categories for Executive Order 12958 and Executive Order 13292 are listed and defined in Table 3-4.

"Originating Agency's Determination Required" or "OADR" should not be used on originally classified technical reports created after

14 October 1995 or originally classified under Executive Order 13292.

If multiple sources are used to derive the classification of the technical report, the latest declassification date applicable to the technical report should be inserted. DoD contractors should use the information specified in the Contract Security Classification Specification or guide furnished with a classified contract or cite the source document. Technical reports containing Restricted Data or Formerly Restricted Data (atomic energy information) should not contain this marking.

**Table 3–4.  Exemption Categories**

| Executive Order 12958 | | Executive Order 13292 | |
|---|---|---|---|
| Exemption Category* | Definition | Exemption Category† | Definition |
| 1 | Reveals an intelligence source, method, or activity or a cryptologic system or activity. | 1 | Reveals the identity of a confidential human source or a human intelligence source or reveals information about the application of an intelligence source or method. |
| 2 | Reveals information that would assist in the development or use of weapons of mass destruction. | 2 | Reveals information that would assist in the development or use of weapons of mass destruction. |
| 3 | Reveals information that would impair the development or use of technology within a U.S. weapons system. | 3 | Reveals information that would impair U.S. cryptologic systems or activities. |
| 4 | Reveals U.S. military plans or national security emergency preparedness plans. | 4 | Reveals information that would impair the application of state of the art technology within a U.S. weapons system. |

*"X1," "X2," etc.
†"25X1," "25X2," etc.

### Table 3-4.    Exemption Categories (Continued)

| Executive Order 12958 | | Executive Order 13292 | |
|---|---|---|---|
| Exemption Category* | Definition | Exemption Category† | Definition |
| 5 | Reveals foreign government information. | 5 | Reveals actual U.S. military war plans that remain in effect. |
| 6 | Damages relations between the United States and a foreign government, reveals a confidential source, or seriously undermines diplomatic activities that are reasonably expected to be ongoing for a period greater than 10 years. | 6 | Reveals information, including foreign government information, that would seriously and demonstrably impair relations between the United States and a foreign government or seriously and demonstrably undermine ongoing diplomatic activities of the United States. |
| 7 | Impairs the ability of responsible U.S. Government officials to protect the President, the Vice President, and other individuals for whom protection services, in the interest of national security, are authorized. | 7 | Reveals information that would clearly and demonstrably impair the current ability of U.S. Government officials to protect the President, the Vice President, and other protectees for whom protection services, in the interest of national security, are authorized. |

*"X1," "X2," etc.
†"25X1," "25X2," etc.

**Table 3-4.    Exemption Categories (Continued)**

| Executive Order 12958 | | Executive Order 13292 | |
| --- | --- | --- | --- |
| Exemption Category* | Definition | Exemption Category† | Definition |
| 8 | Violates a statute, treaty, or international agreement. | 8 | Reveals information that would seriously and demonstrably impair current national security emergency preparedness plans or reveals current vulnerabilities of systems, installations, infrastructures, or projects relating to the national security. |
| | | 9 | Violates a statute, treaty, or international agreement. |

*"X1," "X2," etc.
†"25X1," "25X2," etc.

### 3.12.9    Destruction Notice

Unclassified, limited distribution (distribution statements B, C, D, E, F, or X) technical reports should be handled and destroyed using the same standard as FOUO information; classified technical reports should be handled and destroyed in accordance with DOD 5220.22-M.[41] Unclassified, limited distribution technical reports should not be discarded with recyclable materials.

The following destruction notice should be placed on the front cover and title page of unclassified (except distribution statement A) and classified technical reports: **Destruction Notice: For unclassified, limited distribution documents, destroy by any method that will prevent disclosure of contents or reconstruction of the document For classified documents, destroy in accordance with DOD 5200.1-R, "Information Security Program," Chapter 6, Section C6.7 or DOD 5220.22-M, "National Industrial Security Program Operating Manual," Chapter 5, Section 7.**

In lieu of referencing DOD 5200.1-R, some organizations reference an equivalent DoD component directive/instruction, e.g., Secretary of the Navy Instruction (SECNAVINST) 5510.36.[19]

### 3.12.10    Foreign Government and North Atlantic Treaty Organization (NATO) Information

Classified technical reports that contain classified foreign government information should be marked on the front cover and title page with the following statement if the identity of the country(ies) of origin can be revealed and if intelligence information is not revealed: **This document contains (country[ies] of origin) (indicate classification level) information.** If the country(ies) of origin cannot be revealed, insert **foreign government** in lieu of the country(ies) of origin. If intelligence information is revealed by this statement, the statement should be omitted.

Some foreign governments have a fourth classification level for which there is no U.S. equivalent classification. Table 3-5 lists the foreign classification levels of certain NATO member countries and Australia, Japan, New Zealand, and South Korea and their equivalent U.S. classification; DOD 5200.1-R[26] provides a more extensive list.

If a technical report contains foreign classified information, the classification level should be indicated in English, e.g., "Geheim" would be identified as **GERMAN SECRET**. Foreign government information that falls into the fourth classification level or unclassified foreign government information provided "in confidence" should be identified as such and marked **CONFIDENTIAL - Modified Handling**, e.g., "Diffusion Restreinte" would be identified as follows:
**FRENCH RESTRICTED INFORMATION**
**Protect as CONFIDENTIAL - Modified Handling**.

### Table 3-5.    Foreign Classification Levels

| Country | U.S. Classification Level | | | |
| | Top Secret | Secret | Confidential | Other |
| --- | --- | --- | --- | --- |
| Australia | Top Secret | Secret | Confidential | Restricted |
| Belgium (Flemish) | Zeer Geheim | Geheim | Vertrouwelijk | Bepertke Verspreiding |
| Belgium (French) | Tres Secret | Secret | Confidentiel | Diffusion Restreints |
| Canada | Top Secret | Secret | Confidential | Restricted |
| Denmark | Yderst Hemmeligt | Hemmeligt | Fortroligt | Til Tjenestebrug |
| France | Tres Secret | Secret Defense | Confidentiel | Diffusion Restreinte |

Czech Republic, Greece, Luxembourg, and Turkey not listed in DOD 5200.1-R.[26]
No translation in English characters for Taiwan.

**Table 3-5.    Foreign Classification Levels (Continued)**

| Country | U.S. Classification Level | | | |
|---|---|---|---|---|
| | Top Secret | Secret | Confidential | Other |
| Germany | Streng Geheim | Geheim | VS-Vertraulich | – |
| Hungary | Szigo'uan Titkos | Titkos | Bizalmas | – |
| Iceland | Algjorti | Trunadarmal | – | – |
| Italy | Segretissimo | Segreto | Riservatissimo | Riservato |
| Japan | Kimitsu | Gokuhi | Hi | Toriatsukaichui |
| Korea, South | I Kup Pi Mil | II Kup Pi Mil | III Kup Pi Mil | – |
| Netherlands, The | Zeer Geheim | Geheim | Confidentieel or Vertrouwelijk | Dienstgeheim |
| New Zealand | Top Secret | Secret | Confidential | Restricted |
| Norway | Strengt Hemmelig | Hemmelig | Konfidensiell | Begrenset |
| Poland | Taijny Specjalnego | Tajny | Poufny | – |
| Portugal | Muito Secreto | Secreto | Confidencial | Reservado |
| Spain | Maximo Secreto | Secreto | Confidencial | Diffusion Limitada |
| United Kingdom | Top Secret | Secret | Confidential | Restricted |

Czech Republic, Greece, Luxembourg, and Turkey not listed in DOD 5200.1-R.[26]
No translation in English characters for Taiwan.

Classified technical reports prepared for foreign governments that contain U.S. classified information should be marked as prescribed by the foreign government and should, if the identity of the United States can be revealed and if intelligence information is not revealed, include on the front cover and title page the following statement: **This document contains United States (indicate classification level) information.**

If the technical report contains NATO classified information, the following statement should be included on the front cover and title page: **This document contains NATO (indicate classification level) information.** NATO classification levels are as follows: COSMIC Top Secret, NATO Secret, and NATO Confidential and are equivalent, respectively, to the same U.S. classification levels. When NATO classified information is extracted and inserted in a U.S.-originated technical report, the U.S. originator, for historical and future reference purposes, is the primary office of record and obligated to record and maintain documentation as to what NATO documents and portions are included in the U.S. technical report. This is important in cases of possible compromise, downgrading, declassification, and/or disclosures. Extracted portions should be treated as follows:

a.  The extracted material, i.e., text or visual, should be portion marked.
b.  The page(s) containing NATO classified information should contain the following statement on the top and bottom: **This page contains NATO (indicate classification level) information and should be safeguarded in accordance with USSAN** [United States Security Authority, NATO] **Instruction 1-69 (or appropriate substitute service regulation).**

NATO unclassified information is subject to limited distribution. It should only be released to other NATO staff, to qualified users in the NATO countries, or to non-NATO countries, organizations, or individuals when authorized disclosure is deemed to be in the interest of NATO. NATO unclassified information should not be released to the public.

## 3.13    Seals/Emblems and Logos

The DoD or DoD component seal or emblem, e.g., U.S. Navy emblem, should appear on the front cover and title page of a DoD or DoD component technical report. DoD seals are available at the following Web site: www.defenselink.mil/multimedia/web_graphics.

For a DoD contractor technical report (defined in Section 3.11, item d, page 74), a DoD contractor's logo may appear on the front cover and title page of certain technical reports; however, the DoD or DoD component seal or emblem should not appear in the same technical report. Other seals/emblems or logos are optional; however, these should be kept to a minimum to avoid "cluttering" the front cover and title page. No other graphics, e.g., photographs, illustrations, etc., should appear on the cover (front or back) and title page.

## 3.14    Example Covers and Title Pages

Figures 3-1 to 3-9 show suggested locations for the various items on an unclassified cover (front and back) and title page and a classified cover (front and back) and classified title page, including the inside front cover and the back of the title page. For a contrasting effect, the title information may appear in a serif typeface and the remaining items may appear in a sans serif typeface (shown herein.) The overall classification level should appear in the locations indicated. Items in bold should be prominent.

In Figures 3-1 to 3-9, "DoD" refers to DoD or DoD components; "contractor" refers to DoD contractors (academia and industry). (Figures 3-1 to 3-4 contain the Navy emblem as an example.[m])

## 3.15   Report Documentation Page (Standard Form 298)

The report documentation page or Standard Form 298, as it is commonly known in DoD, is a required element of every defense-related technical report and provides a link between the author and the defense community. It provides the necessary basic information to DTIC so that other researchers are aware of what is occurring in defense-related RDT&E, thus facilitating their research efforts and avoiding unnecessary and costly duplication of work. The Standard Form 298 also enables other organizational libraries or TICs to use the same form and information for their own STI databases.

The Standard Form 298 is a two-sided document. Figures 3-10 and 3-11 are modified re-creations of the front and back, respectively, of a Standard Form 298. Only the front (Figure 3-10) should appear in a technical report, and it should be numbered page iii. The Standard Form 298 should not be included in the table of contents.

The classification of the Standard Form 298 is addressed in item 16 on the form; therefore, it is unnecessary to place other classification markings on the page in a classified technical report. The back (as it appears in the technical report) of the Standard Form 298 is blank in unclassified and classified technical reports and serves as page iv.

Completing a Standard Form 298 is estimated to average about 1 hour. Most of the information can be obtained from the front cover or title page of the technical report, which should expedite the process. This includes the following, which pertains to the numbered instructions for completing a Standard Form 298 (Figure 3-11):

a.   Report (publication) date (instruction 1). The Standard Form 298 indicates day, month, and year and stipulates that, at a minimum, the year of the technical report must be indicated; however, to conform with Section 3.8, the month and year should be provided.

b.   Report type (instruction 2). The type of report should be "technical." "Final" and "interim" may be included, as appropriate.

c.   Dates covered (instruction 3). These dates indicate when the work was performed, not when the technical report is published.

[m]Permission granted to use the Navy emblem by the Department of the Navy (www.chinfo.navy.mil/navpalib/questions/graphics.html). Neither the Department of the Navy nor any other component of the Department of Defense has approved, endorsed, or authorized this book.

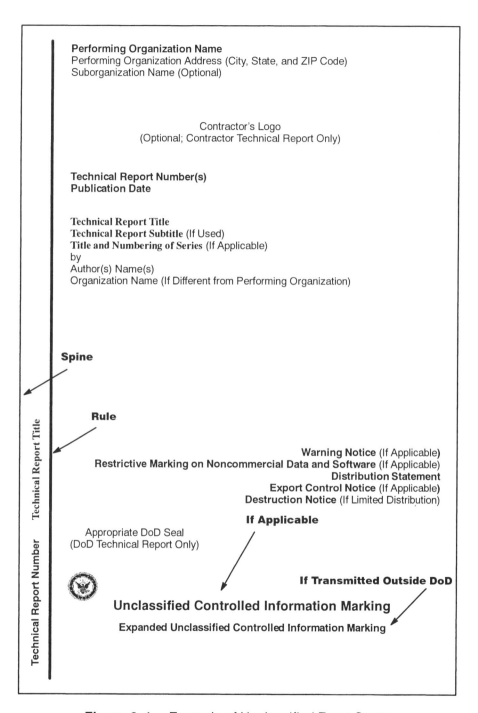

**Figure 3-1.**    Example of Unclassified Front Cover

**Performing Organization Name**
Performing Organization Address (City, State, and ZIP Code)
Suborganization Name (Optional)

Contractor's Logo
(Optional; Contractor Technical Report Only)

**Technical Report Number(s)**
**Publication Date**

**Technical Report Title**
**Technical Report Subtitle** (If Used)
**Title and Numbering of Series** (If Applicable)
by
Author(s) Name(s)
Organization Name (If Different from Performing Organization)
Organization Address (If Different from Performing Organization)
DoD Contract/Grant Number (Contractor Technical Report Only)

**Warning Notice** (If Applicable)
**Restrictive Marking on Noncommercial Data and Software** (If Applicable)
**Distribution Statement**
**Export Control Notice** (If Applicable)
**Destruction Notice** (If Limited Distribution)

**If Applicable**

Appropriate DoD Seal
(DoD Technical Report Only)

**If Transmitted Outside DoD**

**Unclassified Controlled Information Marking**

**Expanded Unclassified Controlled Information Marking**

**Figure 3-2.**   Example of Unclassified Title Page

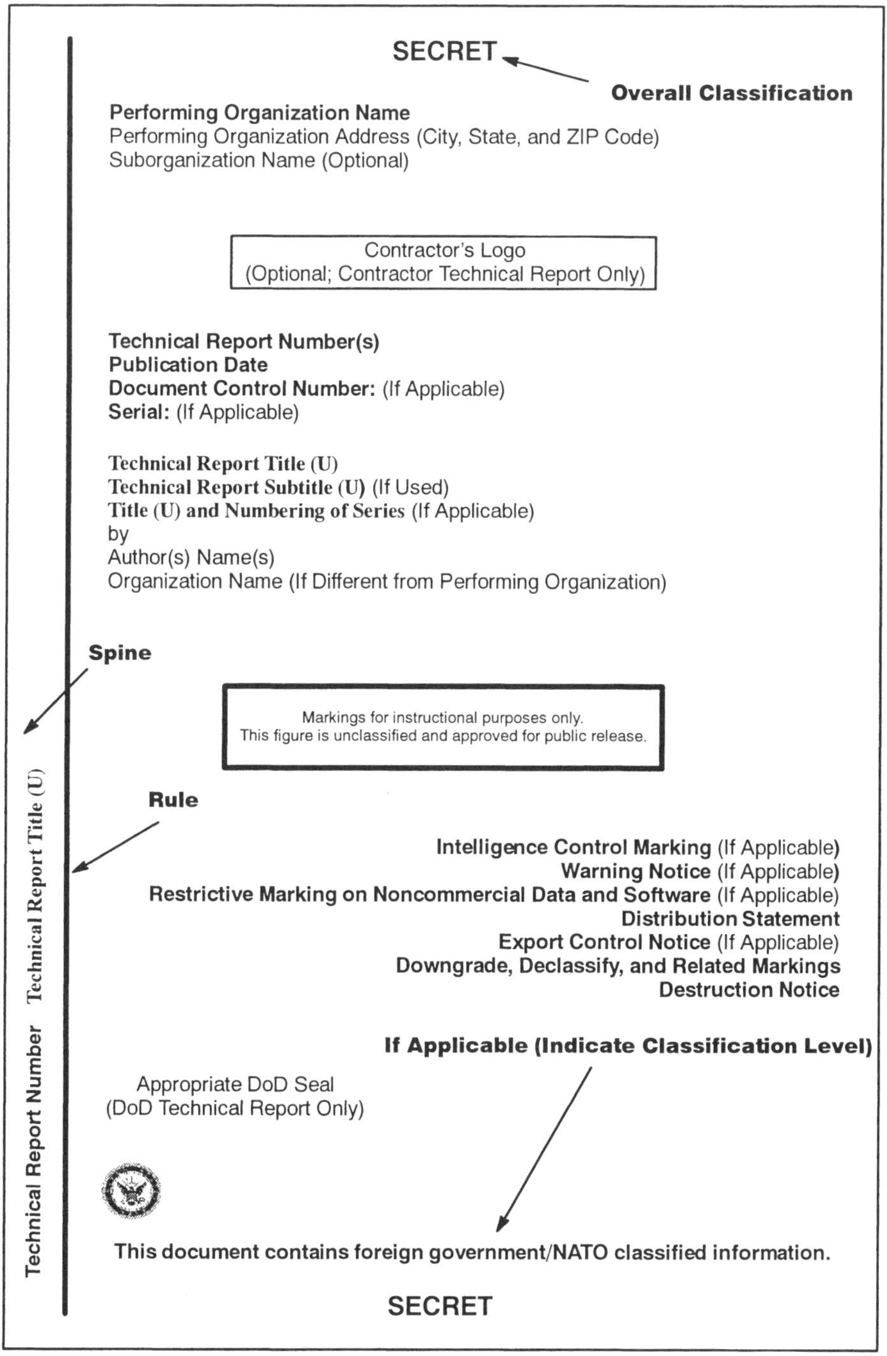

SECRET ◄—

**Overall Classification**

**Performing Organization Name**
Performing Organization Address (City, State, and ZIP Code)
Suborganization Name (Optional)

Contractor's Logo
(Optional; Contractor Technical Report Only)

**Technical Report Number(s)**
**Publication Date**
**Document Control Number:** (If Applicable)
**Serial:** (If Applicable)

**Technical Report Title (U)**
**Technical Report Subtitle (U)** (If Used)
**Title (U) and Numbering of Series** (If Applicable)
by
Author(s) Name(s)
Organization Name (If Different from Performing Organization)

**Spine**

Markings for instructional purposes only.
This figure is unclassified and approved for public release.

**Rule**

**Intelligence Control Marking** (If Applicable)
**Warning Notice** (If Applicable)
**Restrictive Marking on Noncommercial Data and Software** (If Applicable)
**Distribution Statement**
**Export Control Notice** (If Applicable)
**Downgrade, Declassify, and Related Markings**
**Destruction Notice**

**If Applicable (Indicate Classification Level)**

Appropriate DoD Seal
(DoD Technical Report Only)

**This document contains foreign government/NATO classified information.**

SECRET

*Technical Report Number    Technical Report Title (U)*

**Figure 3-3.** Example of Classified Front Cover

SECRET

**Overall Classification**

**Performing Organization Name**
Performing Organization Address (City, State, and ZIP Code)
Suborganization Name

Contractor's Logo
(Optional; Contractor Technical Report Only)

**Technical Report Number(s)**
**Publication Date**
**Document Control Number:** (If Applicable)
**Serial:** (If Applicable)

**Technical Report Title (U)**
**Technical Report Subtitle (U)** (If Used)
**Title (U) and Numbering of Series** (If Applicable)
by
Author(s) Name(s)
Organization Name (If Different from Performing Organization)
Organization Address (If Different from Performing Organization)
DoD Contract/Grant Number (Contractor Technical Report Only)

Markings for instructional purposes only.
This figure is unclassified and approved for public release.

**Intelligence Control Marking** (If Applicable)
**Warning Notice** (If Applicable)
**Restrictive Marking on Noncommercial Data and Software** (If Applicable)
**Distribution Statement**
**Export Control Notice** (If Applicable)
**Downgrade, Declassify, and Related Markings**
**Destruction Notice**

**If Applicable (Indicate Classification Level)**

Appropriate DoD Seal
(DoD Technical Report Only)

**This document contains foreign government/NATO classified information.**

SECRET

**Figure 3-4.** Example of Classified Title Page

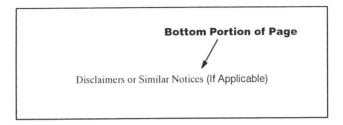

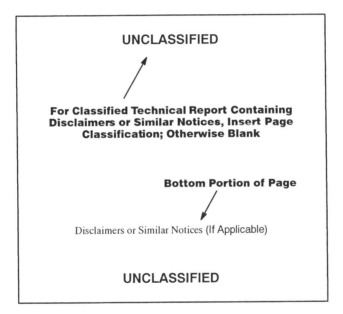

**Inside Front Cover (Classified Technical Report)**

**Figure 3-5.**    Examples of Inside Front Cover

d.  Title (instruction 4). The title and subtitle, if used, should be copied verbatim from the front cover or title page. A short (abbreviated) title should not be used. If the technical report is classified, the classification of the title and subtitle, if used, is part of the title and subtitle and should be indicated as such, e.g., "(U)."

e.  Contract number, grant number, program element number, project number, task number, and work unit number (instruction 5). Not all

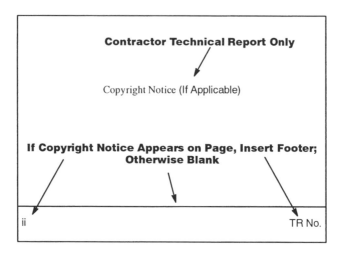

**Figure 3-6.**   Example of Back of Title Page
(Unclassified Technical Report)

of these may be applicable; however, if used, they should match the same information in the preface (administrative information) (Section 3.20) of the technical report.

f.   Author(s) (instruction 6). Unlike the front cover and title page, the form of entry is last name, first name, middle initial, and additional qualifiers separated by commas, e.g., Smith, Richard, J., Jr.

g.   Performing organization name(s) and address(es) (instruction 7). This may be taken from the front cover or title page; however, the street address should be included with the city, State, and ZIP code.

h.   Performing organization report number (instruction 8). This should be copied verbatim from the front cover or title page.

i.   Sponsoring/monitoring agency name(s) and addresses (instruction 9). The name and address (including street address) of the organization(s) financially responsible for and monitoring the work should be entered. For DoD technical reports, an organization such as NAV-SEA may sponsor or monitor the work of a suborganization, such as NSWCCD. Normally, a DoD contractor technical report is sponsored or monitored by a DoD component organization. However, as an exception (as the case of in-house-funded IR&D), the same information from item g may be entered.

j.   Sponsor/monitor's acronym(s) (instruction 10). For example, NAV-SEASYSCOM (complete acronym) should be inserted for the Naval Sea Systems Command.

k.   Sponsor/monitor's report number(s) (instruction 11). If the sponsor/ monitor has a separate technical report number from the performing

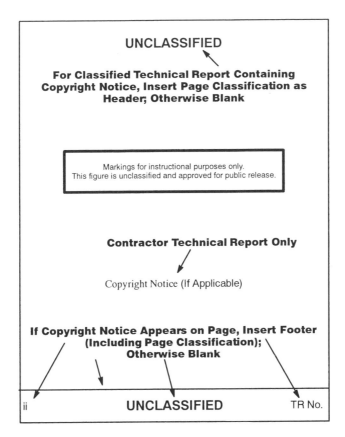

**Figure 3-7.**    Example of Back of Title Page (Classified
Technical Report)

organization technical report number, it should be provided; other-wise, the performing organization technical report number should be repeated.

l.  Distribution/availability statement (instruction 12). Agency-man-dated availability statements should be used to indicate the public availability or distribution limitations of the technical report. The distribution statement should be copied verbatim from the front cov-er or title page. If additional limitations/restrictions or special mark-ings are indicated, such as intelligence control markings, agency au-thorization procedures should be followed. Copyright information from page ii should be included, if applicable. The following abbre-viations for intelligence control markings should be used:

(1)  "REL TO (abbreviated name of country or organization)" should be used for "Authorized for Release to (name of country

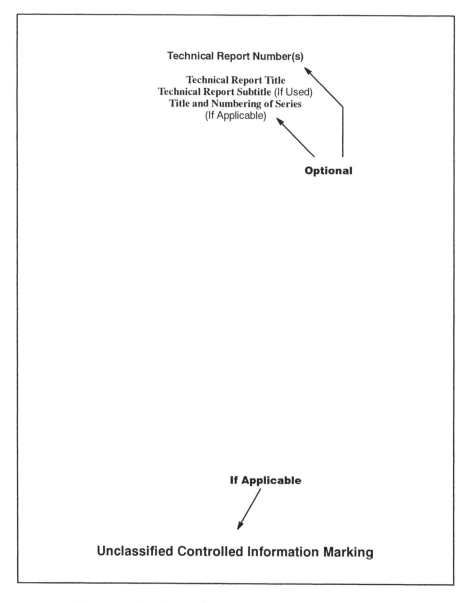

**Figure 3-8.**    Example of Back Cover of Unclassified
Technical Report

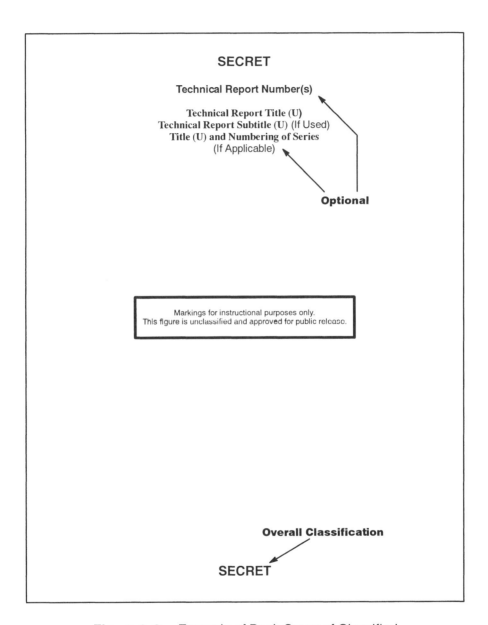

SECRET

Technical Report Number(s)

Technical Report Title (U)
Technical Report Subtitle (U) (If Used)
Title (U) and Numbering of Series
(If Applicable)

**Optional**

Markings for instructional purposes only.
This figure is unclassified and approved for public release.

**Overall Classification**

SECRET

**Figure 3-9.**   Example of Back Cover of Classified
Technical Report

| REPORT DOCUMENTATION PAGE | Form Approved OMB No. 0704-0188 |
|---|---|

The public reporting burden for this collection of information is estimated to average 1 hour per response, including the time for reviewing instructions, searching existing data sources, gathering and maintaining the data needed, and completing and reviewing the collection of information. Send comments regarding this burden estimate or any other aspect of this collection of information, including suggestions for reducing the burden, to Department of Defense, Washington Headquarters Services, Directorate for Information Operations and Reports (0704-0188), 1215 Jefferson Davis Highway, Suite 1204, Arlington, VA 22202-4302. Respondents should be aware that notwithstanding any other provision of law, no person shall be subject to any penalty for failing to comply with a collection of information if it does not display a currently valid OMB control number.
**PLEASE DO NOT RETURN YOUR FORM TO THE ABOVE ADDRESS**

| 1. REPORT DATE (DD-MM-YYYY) | 2. REPORT TYPE | 3. DATES COVERED (From - To) |
|---|---|---|

| 4. TITLE AND SUBTITLE | 5a. CONTRACT NUMBER |
|---|---|
| | 5b. GRANT NUMBER |
| | 5b. PROGRAM ELEMENT NUMBER |
| 6. AUTHOR(S) | 5d. PROJECT NUMBER |
| | 5e. TASK NUMBER |
| | 5f. WORK UNIT NUMBER |

| 7.PERFORMING ORGANIZATION NAME(S) AND ADDRESS(ES) | 8. PERFORMING ORGANIZATION REPORT NUMBER |
|---|---|

| 9. SPONSORING/MONITORING AGENCY NAME(S) AND ADDRESS(ES) | 10. SPONSOR/MONITOR'S ACRONYM(S) |
|---|---|
| | 11. SPONSOR/MONITOR'S REPORT NUMBER(S) |

12.. DISTRIBUTION/AVAILABILITY STATEMENT

13. SUPPLEMENTARY NOTES

14. ABSTRACT

15. SUBJECT TERMS

| 16. SECURITY CLASSIFICATION OF: | | | 17. LIMITATION OF ABSTRACT | 18. NUMBER OF PAGES | 19a. NAME OF RESPONSIBLE PERSON |
|---|---|---|---|---|---|
| a. REPORT | b. ABSTRACT | c. THIS PAGE | | | 19b. TELEPHONE NUMBER (Include area code) |

**Page Number** ⟶ iii

Standard Form 298 (Rev. 8/98)
Prescribed by ANSI Std. Z39.18

**Figure 3-10.**   Standard Form 298 (Front)

## INSTRUCTIONS FOR COMPLETING SF 298

**1. REPORT DATE.** Full publication date, including day, month, if available. Must cite at least the year and be Year 2000 compliant, e.g. 30-06-1998; xx-06-1998; xx-xx-1998.

**2. REPORT TYPE.** State the type of report, such as final, technical, interim, memorandum, master's thesis, progress, quarterly, research, special, group study, etc.

**3. DATES COVERED.** Indicate the time during which the work was performed and the report was written, e.g., Jun 1997 - Jun 1998; 1-10 Jun 1996; May - Nov 1998; Nov 1998.

**4. TITLE.** Enter title and subtitle with volume number and part number, if applicable. On classified documents, enter the title classification in parentheses.

**5a. CONTRACT NUMBER.** Enter all contract numbers as they appear in the report, e.g. F33615-86-C-5169.

**5b. GRANT NUMBER.** Enter all grant numbers as they appear in the report, e.g. AFOSR-82-1234.

**5c. PROGRAM ELEMENT NUMBER.** Enter all program element numbers as they appear in the report, e.g. 61101A.

**5d. PROJECT NUMBER.** Enter all project numbers as they appear in the report, e.g. 1F665702D1257; ILIR.

**5e. TASK NUMBER.** Enter all task numbers as they appear in the report, e.g. 05; RF0330201; T4112.

**5f. WORK UNIT NUMBER.** Enter all work unit numbers as they appear in the report, e.g. 001; AFAPL30480105.

**6. AUTHOR(S).** Enter name(s) of person(s) responsible for writing the report, performing the research, or credited with the content of the report. The form of entry is the last name, first name, middle initial, and additional qualifiers separated by commas, e.g. Smith, Richard, J, Jr.

**7. PERFORMING ORGANIZATION NAME(S) AND ADDRESS(ES).** Self-explanatory.

**8. PERFORMING ORGANIZATION REPORT NUMBER.** Enter all unique alphanumeric report numbers assigned by the performing organization, e.g. BRL-1234; AFWL-TR-85-4017-Vol-21-PT-2.

**9. SPONSORING/MONITORING AGENCY NAME(S) AND ADDRESS(ES).** Enter the name and address of the organization(s) financially responsible for and monitoring the work.

**10. SPONSOR/MONITOR'S ACRONYM(S).** Enter, if available, e.g. BRL, ARDEC, NADC.

**11. SPONSOR/MONITOR'S REPORT NUMBER(S).** Enter report number as assigned by the sponsoring/monitoring agency, if available, e.g. BRL-TR-829; -215.

**12. DISTRIBUTION/AVAILABILITY STATEMENT.** Use agency-mandated availability statements to indicate the public availability or distribution limitations of the report. If additional limitations/restrictions or special markings are indicated, follow agency authorization procedures, e.g. RD/FRD, PROPIN, ITAR, etc. Include copyright information.

**13. SUPPLEMENTARY NOTES.** Enter information not included elsewhere such as: prepared in cooperation with; translation of; report supersedes; old edition number, etc.

**14. ABSTRACT.** A brief (approximately 200 words) factual summary of the most significant information.

**15. SUBJECT TERMS.** Key words or phrases identifying major concepts in the report.

**16. SECURITY CLASSIFICATION.** Enter security classification in accordance with security classification regulations, e.g. U, C, S, etc. If this form contains classified information, stamp classification level on the top and bottom of this page.

**17. LIMITATION OF ABSTRACT.** This block must be completed to assign a distribution limitation to the abstract. Enter UU (Unclassified Unlimited) or SAR (Same as Report). An entry in this block is necessary if the abstract is to be limited.

Standard Form 298 Back (Rev. 8/98)

**Figure 3-11.** Standard Form 298 (Back)

or international organization)." Table 3-6 lists the abbreviations[n] of NATO member countries and Australia, Japan, New Zealand, Taiwan, and South Korea.

**Table 3-6.  Foreign Country Abbreviations**

| Country | Abbreviation |
|---|---|
| Australia | AS |
| Belgium | BE |
| Canada | CA |
| Czech Republic | EZ |
| Denmark | DA |
| France | FR |
| Germany | GM |
| Greece | GR |
| Hungary | HU |
| Iceland | IC |
| Italy | IT |
| Japan | JA |
| Korea, South | KS |
| Luxembourg | LU |
| Netherlands, The | NL |
| New Zealand | NZ |
| Norway | NO |
| Poland | PL |
| Portugal | PO |
| Spain | SP |
| Taiwan | TW |
| Turkey | TU |
| United Kingdom | UK |

[n]The abbreviations were obtained from FIPS 10-4, which is maintained by the Department of State and published by the National Institute of Standards and Technology. FIPS 10-4 codes are intended for general use throughout the Government, especially in activities associated with the mission of the Department of State and national defense programs.

(2) "PROPIN" should be used for "Caution—Proprietary Information Involved."

(3) "ORCON" should be used for "Dissemination and Extraction of Information Controlled by Originator."

(4) "NOCONTRACT" should be used for "Not Releasable to Contractors/Consultants."

(5) "NOFORN" should be used for "Not Releasable to Foreign Nationals."

Table 3-7 lists the abbreviations for warning notices.

**Table 3-7.  Warning Notice Abbreviations**

| Notice | Abbreviation |
|---|---|
| COMSEC Material—Access by contractor personnel restricted to U.S. citizens holding final Government clearance. | None |
| Critical Nuclear Weapon Design Information—DoD Directive 5210.2 applies. | N |
| Reproduction requires approval of originator or higher DoD authority. | None |
| Further dissemination only as directed by (insert appropriate command or official) or higher DoD authority. | None |
| Special Handling Required—Not releasable to foreign nationals. | None |
| This document (or material) is subject to special export controls, and each transmittal to foreign governments or foreign nationals may be made only with prior approval of (originating command). | None |
| RESTRICTED DATA—This material contains Restricted Data as defined in the Atomic Energy Act of 1954. Unauthorized disclosure is subject to administrative and criminal sanctions. | RD |

**Table 3-7. Warning Notice Abbreviations (Continued)**

| Notice | Abbreviation |
|---|---|
| FORMERLY RESTRICTED DATA—Unauthorized disclosure subject to administrative and criminal sanctions. Handle as Restricted Data in foreign dissemination. Section 144.b, Atomic Energy Act, 1954. | FRD |
| WARNING—This document contains technical data whose export is restricted by the Arms Export Control Act (Title 22 U.S.C., Sec. 2751 et seq.) or the Export Administration Act of 1979, as amended (Title 50 U.S.C., App. 2401 et seq.). Violations of these export laws are subject to severe criminal penalties. Disseminate in accordance with the provisions of DoD Directive 5230.25. | None |

    m. Supplementary notes (instruction 13). Information not included elsewhere should be entered, such as "prepared in cooperation with," "translation of," "technical report supersedes," "old edition," etc.

    n. Abstract (instruction 14). The abstract should be copied verbatim from the technical report, including the classification of the abstract in classified technical reports. Abstracts are discussed in detail in Section 3.16. "Keywords," which accompany abstracts, are addressed in item o as "subject terms."

    o. Subject terms (instruction 15). The keywords or phrases following the abstract should be copied verbatim and inserted here.

    p. Security classification (instruction 16). The security classification of the technical report, the abstract, and the information on the Standard Form 298 ("this page"), respectively, should be entered as follows:"TS" for "Top Secret," "S" for "Secret, "C" for Confidential, and "U" for unclassified. If the technical report is unclassified, "U" should be entered in all three spaces.

    q. Limitation of abstract (instruction 17). This block should be completed to assign a distribution limitation to the abstract. Either "UU" (for "unclassified, unlimited" distribution) or "SAR" (for "Same as Report") should be entered. If the technical report is assigned distribution statement A (unclassified; approved for public release), UU should be entered; otherwise, SAR should be entered.

ANSI/NISO Z39.18-2005 states that a report documentation page "is an optional component for academic and industrial [technical] reports."[2] This statement does not apply to defense-related technical reports prepared by DoD contractors, which require a report documentation page.

A sample completed Standard Form 298 appears as Appendix E in ANSI/NISO Z39.18-2005; a blank Standard Form 298 (with the instructions for completing it) is available from DTIC at the following Web site: http://www.dtic.mil/dtic/forms/SF0298_fillable.pdf. This form may be completed online and downloaded.

## 3.16 Abstract

An abstract presents a concise (approximately 200 words . . . ) informative statement of the purpose, scope, methods, and major findings of the [technical] report, including results, conclusions, and recommendations. The informative abstract retains the tone and scope of the [technical] report but omits the details.[2]

An abstract should consist of only one paragraph and should not contain any figures or tables. An abstract may contain unclassified controlled information; however, it should be unclassified in a classified technical report to allow its inclusion in unclassified databases.

As stated earlier, ANSI/NISO Z39.14-1997 is the national standard for abstracts and is referenced in ANSI/NISO Z39.18-2005. ANSI/NISO Z39.14-1997 further clarifies the definition of an abstract by stating the following: "The term *abstract* should not be confused with the related but distinct terms: *annotation, extract, summary,* and *synoptic*."[12] This is noteworthy because some authors feel that an abstract suffices for an executive summary and vice versa. This is not true: Both have a related but separate purpose. ANSI/NISO Z39.14-1997 adds the following:

A well-prepared abstract enables readers (a) to identify the basic content of a document [technical report] quickly, (b) to determine its relevance to their interests, and thus (c) to decide whether they need to read the document in its entirety. The abstract may facilitate a closer reading of the primary document [body of the technical report] by providing an introductory overview of its topic or argument, or, for readers to whom the document is of marginal interest, the abstract may provide enough information to make a reading of the full document unnecessary.[12]

While having a separate purpose, the preceding description of an abstract parallels the description of an executive summary, which is discussed in Chapter 4, Section 4.1. The executive summary can serve as a foundation for the abstract; thus, the abstract should be written after the executive summary. Some authors also feel that a condensed version of the introduction to the technical report suffices for an abstract. This is incorrect: The abstract is a summary of the entire technical report.

ANSI/NISO Z39.18-2005 also states that "an abstract contains no undefined symbols, abbreviations, or acronyms and makes no reference by number to references or illustrative material."[2] ANSI/NISO Z39.14-1997 is more explicit: "An abstract must be intelligible to a reader without reference to the document [technical report] it represents. For clarity, avoid using footnotes, lists of references, or references to the text of the original document."[12] The last sentence precludes brief descriptions of the elements of the main body of the technical report, e.g., "Section 1 is an introduction."

Defense-related technical reports, by their nature, contain many acronyms, and it is important that they be identified in the abstract. As in the title and subtitle of the technical report, acronyms should be avoided in the abstract. They should only be used if a term appears more than once in the abstract or if the acronym is more commonly known than its spelled-out equivalent. In either case, the acronym should appear in parentheses after its spelled-out equivalent after its first mention.

The abstract is the first of several standalone elements within the overall technical report, the others being the table of contents, the lists of figures and tables, the executive summary, and, to a certain degree, the figures and tables, themselves. As a standalone element and as stated earlier, the abstract should be able to be understood independent of references to other elements of the technical report. A common mistake of many authors is to refer to references or appendixes in the abstract by, simply, their number or letter. Citing references or appendixes should be avoided altogether, but if it is necessary, a brief description of their contents is all that is necessary.

There are two types of abstracts: informative and indicative.[12] Technical report abstracts should be informative and "state the purpose, methodology, results, and conclusion presented in the original document [technical report]."[12]

A minimum of three keywords or phrases pertaining to the subject matter of the technical report should accompany the abstract; these keywords should be included on the Standard Form 298 (Section 3.15). The keywords are used not only by DTIC but other libraries or TICs to catalog and index the technical report. The keywords may be taken from the title and subtitle and, if necessary, from the abstract, itself. In classified technical reports, the keywords or phrases, like the abstract, itself, should be unclassified to facilitate cataloging and indexing. The keywords should be placed after or underneath the abstract in alphabetical order. An acronym, alone, is not a keyword; instead, it should be spelled out and placed in parentheses after its spelled-out equivalent.

The abstract should be written in the third person and in the present and past tense. The portion of the abstract that refers to the introduction and the recommendations elements in the technical report should be written in the present tense; the rest of the abstract should be written in the past tense.

Figures 3–12 and 3–13 show examples of an abstract in an unclassified and classified technical report, respectively, using a typographical progression format or a decimal numbering format.

The abstract should appear on page v of the front matter.

## 3.17   Table of Contents

ANSI/NISO Z39.18-2005 states the following regarding the contents section, commonly referred to as the table of contents:

> The required [table of] contents section identifies the heading and location of . . . each major section of the front matter (excluding the title page and the contents section itself), the content [body], and the back matter. A contents section helps readers understand the organization and scope of a [technical] report. Headings in a table of contents are worded, spelled, punctuated, and numbered, if used, exactly as they are in the [technical] report.

> It is useful to include a list of subheadings in the [table of] contents section at the beginning of each major [technical] report section that is more than 20 pages in length. Subheadings are also helpful for understanding complex material; however, not all levels of headings need to be listed in the contents section. First- and second-level headings may suffice.[2]

While ANSI/NISO Z39.18-2005 states that the table of contents is spelled exactly as the headings and subheadings appear in the technical report, it is unnecessary to include ancillary parenthetical information sometimes accompanying a heading or subheading. It is not uncommon for defense-related technical reports to include three, and possibly four, heading levels in the table of contents, regardless of the number of pages in a technical report element or section.

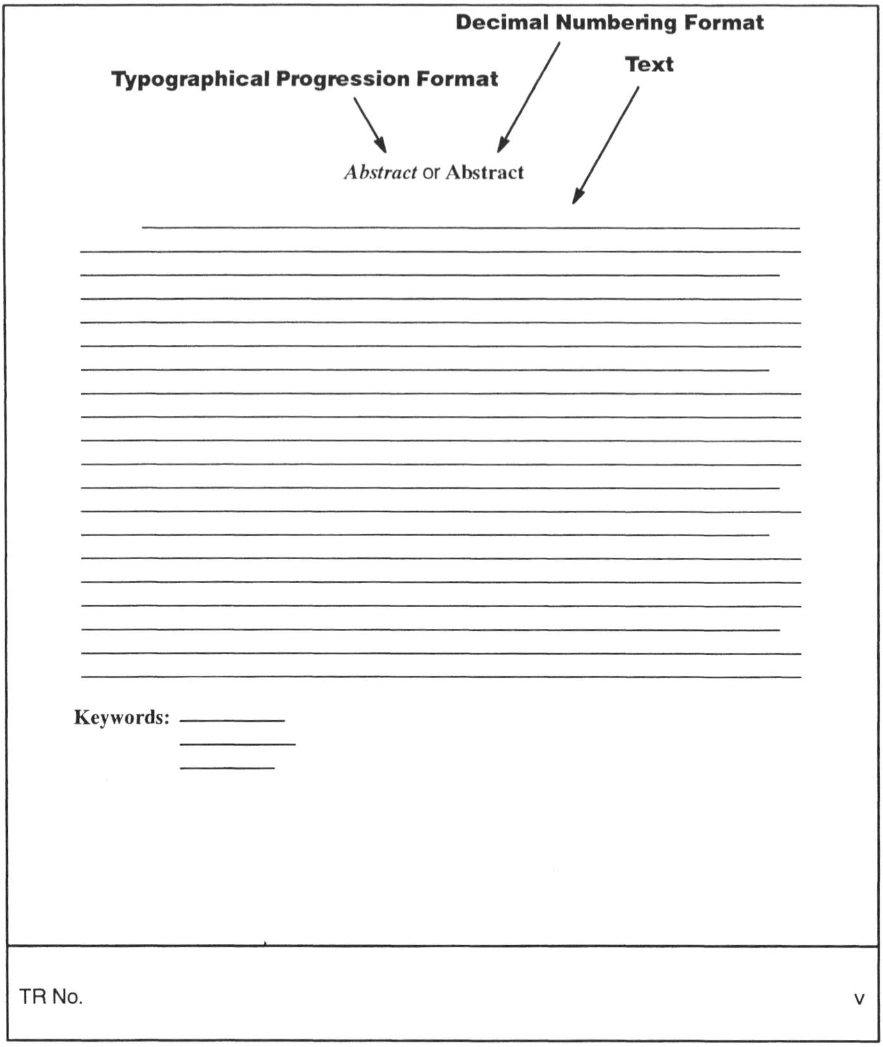

**Decimal Numbering Format**

**Typographical Progression Format**

**Text**

*Abstract* or Abstract

Keywords: ─────

TR No.                                                                          v

**Figure 3-12.**   Example of Abstract in Unclassified
Technical Report

In addition to the guidance presented by ANSI/NISO Z39.18-2005, the following should be observed regarding headings and subheadings and the table of contents:

- The table of contents should be titled, simply, "Contents"—not "Contents Section" or "Table of Contents"—in the technical report.
- Generally, headings should be different (unique). A lower-level heading should be different than a higher-level heading. Headings should also be different at the same level; however, two or more subhead-

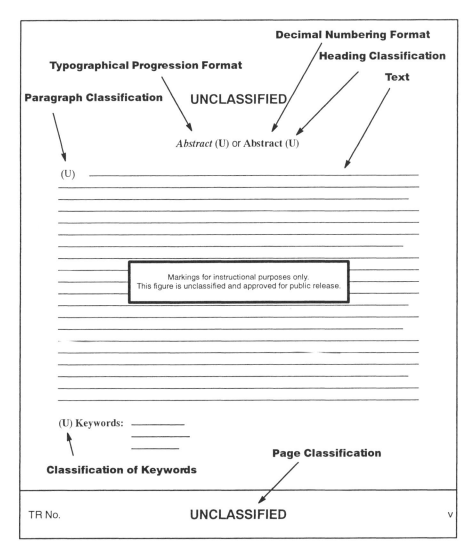

Decimal Numbering Format

Heading Classification

Text

Typographical Progression Format

Paragraph Classification

UNCLASSIFIED

*Abstract* (U) or Abstract (U)

(U) ————————————————————————

Markings for instructional purposes only.
This figure is unclassified and approved for public release.

(U) Keywords: ————

Classification of Keywords

Page Classification

TR No.                    UNCLASSIFIED                    v

**Figure 3-13.**    Example of Abstract in Classified
Technical Report

ings may be identical if they appear under different headings. This is
most easily noticed when reviewing the table of contents. Figure 3-14 shows examples of what is incorrect and what is correct
with regard to heading titles at the same level. (Affected headings are
highlighted in boldface.)

• The table of contents should have at least two entries for each level
of heading. Some authors mistakenly have only a single subheading
under a higher-level heading. This, too, is most easily noticed when

**Incorrect**

Methods, Assumptions, and Procedures (First, Second, and Third Tests) . . . . . . . . . . . . . . 7
  **Methods** . . . . . . . . . . . . . . . . . . . . . . . . . . . . . . . . . . . . . . . . . . . . . . . . . . . . . . . . . . . . . . . . 8
  **Assumptions** . . . . . . . . . . . . . . . . . . . . . . . . . . . . . . . . . . . . . . . . . . . . . . . . . . . . . . . . . . . . 9
  **Procedures** . . . . . . . . . . . . . . . . . . . . . . . . . . . . . . . . . . . . . . . . . . . . . . . . . . . . . . . . . . . . . 10
    First Procedure . . . . . . . . . . . . . . . . . . . . . . . . . . . . . . . . . . . . . . . . . . . . . . . . . . . . . . 11
    Second Procedure . . . . . . . . . . . . . . . . . . . . . . . . . . . . . . . . . . . . . . . . . . . . . . . . . . . 12
    Third Procedure . . . . . . . . . . . . . . . . . . . . . . . . . . . . . . . . . . . . . . . . . . . . . . . . . . . . 13
  **Methods** . . . . . . . . . . . . . . . . . . . . . . . . . . . . . . . . . . . . . . . . . . . . . . . . . . . . . . . . . . . . . . . 14
  **Assumptions** ◄ - - - - - - - - - **Identical Headings at Same Level** · · · · 15
  **Procedures** . . . . . . . . . . . . . . . . . . . . . . . . . . . . . . . . . . . . . . . . . . . . . . . . . . . . . . . . . . . . . 16
    First Procedure . . . . . . . . . . . . . . . . . . . . . . . . . . . . . . . . . . . . . . . . . . . . . . . . . . . . . . 17
    Second Procedure . . . . . . . . . . . . . . . . . . . . . . . . . . . . . . . . . . . . . . . . . . . . . . . . . . . 18
    Third Procedure . . . . . . . . . . . . . . . . . . . . . . . . . . . . . . . . . . . . . . . . . . . . . . . . . . . . 19
  **Methods** . . . . . . . . . . . . . . . . . . . . . . . . . . . . . . . . . . . . . . . . . . . . . . . . . . . . . . . . . . . . . . . 20
  **Assumptions** . . . . . . . . . . . . . . . . . . . . . . . . . . . . . . . . . . . . . . . . . . . . . . . . . . . . . . . . . . . . 21
  **Procedures** . . . . . . . . . . . . . . . . . . . . . . . . . . . . . . . . . . . . . . . . . . . . . . . . . . . . . . . . . . . . . 22
    First Procedure . . . . . . . . . . . . . . . . . . . . . . . . . . . . . . . . . . . . . . . . . . . . . . . . . . . . . . 23
    Second Procedure . . . . . . . . . . . . . . . . . . . . . . . . . . . . . . . . . . . . . . . . . . . . . . . . . . . 24
    Third Procedure . . . . . . . . . . . . . . . . . . . . . . . . . . . . . . . . . . . . . . . . . . . . . . . . . . . . 25

**Correct**

Methods, Assumptions, and Procedures . . . . . . . . . . . . . . . . . . . . . . . . . . . . . . . . . . . . . 7
  **Methods (First Test)** . . . . . . . . . . . . . . . . . . . . . . . . . . . . . . . . . . . . . . . . . . . . . . . . . . . . 8
  **Assumptions (First Test)** . . . . . . . . . . . . . . . . . . . . . . . . . . . . . . . . . . . . . . . . . . . . . . . . 9
  **Procedures (First Test)** . . . . . . . . . . . . . . . . . . . . . . . . . . . . . . . . . . . . . . . . . . . . . . . . . 10
    First Procedure . . . . . . . . . . . . . . . . . . . . . . . . . . . . . . . . . . . . . . . . . . . . . . . . . . . . . . 11
    Second Procedure . . . . . . . . . . . . . . . . . . . . . . . . . . . . . . . . . . . . . . . . . . . . . . . . . . . 12
    Third Procedure . . . . . . . . . . . . . . . . . . . . . . . . . . . . . . . . . . . . . . . . . . . . . . . . . . . . 13
  **Methods (Second Test)** . . . . . . . . . . . . . . . . . . . . . . . . . . . . . . . . . . . . . . . . . . . . . . . . 14
  **Assumptions (Second Test)** ◄ - - - **Different Headings at Same Level** . 15
  **Procedures (Second Test)** . . . . . . . . . . . . . . . . . . . . . . . . . . . . . . . . . . . . . . . . . . . . . . 16
    First Procedure . . . . . . . . . . . . . . . . . . . . . . . . . . . . . . . . . . . . . . . . . . . . . . . . . . . . . . 17
    Second Procedure . . . . . . . . . . . . . . . . . . . . . . . . . . . . . . . . . . . . . . . . . . . . . . . . . . . 18
    Third Procedure . . . . . . . . . . . . . . . . . . . . . . . . . . . . . . . . . . . . . . . . . . . . . . . . . . . . 19
  **Methods (Third Test)** . . . . . . . . . . . . . . . . . . . . . . . . . . . . . . . . . . . . . . . . . . . . . . . . . . 20
  **Assumptions (Third Test)** . . . . . . . . . . . . . . . . . . . . . . . . . . . . . . . . . . . . . . . . . . . . . . . 21
  **Procedures (Third Test)** . . . . . . . . . . . . . . . . . . . . . . . . . . . . . . . . . . . . . . . . . . . . . . . . 22
    First Procedure . . . . . . . . . . . . . . . . . . . . . . . . . . . . . . . . . . . . . . . . . . . . . . . . . . . . . . 23
    Second Procedure . . . . . . . . . . . . . . . . . . . . . . . . . . . . . . . . . . . . . . . . . . . . . . . . . . . 24
    Third Procedure . . . . . . . . . . . . . . . . . . . . . . . . . . . . . . . . . . . . . . . . . . . . . . . . . . . . 25

Figure 3–14.   Headings at Same Level

reviewing the table of contents. If it appears, it can be remedied by combining the single subheading with the preceding higher-level heading, by deleting the subheading altogether, or by splitting the subheading into two subheadings. Figure 3-15 shows examples of what is incorrect and what is correct with regard to the minimum number of headings. (Affected headings are highlighted in boldface.)

- Acronyms should not appear by themselves in a heading or subheading, if at all, regardless if they have been identified previously in the technical report. Like the abstract and executive summary, a table of contents is a standalone element in the technical report that is often reviewed by a reader before reading the rest of the technical report; thus, an acronym, if part of a heading or subheading, should be placed in parentheses and preceded by its spelled-out equivalent.

- If the executive summary contains headings and subheadings, these should be included in the table of contents of the technical report. (Headings and subheadings may be repeated verbatim from the main body of the technical report.)

- Appendix titles should be included in the table of contents of the technical report. If an appendix contains headings and subheadings, these may optionally be included in the table of contents of the technical report (not shown in the examples herein); however, if an appendix contains its own table of contents, e.g., a separate document, its headings and subheadings should not be repeated in the table of contents of the technical report. Also, for consistency in the table of contents of the technical report, if a technical report contains more than one appendix and one of the appendixes contains its own table of contents, then none of the appendix headings and subheadings from the other appendixes should appear in the table of contents of the technical report.

- A minimum of three leaders (dots across the page) should appear between an entry and its corresponding page number. Runover lines should be indented.

- In classified technical reports, the table of contents should, if possible, be unclassified. However, the table of contents needs to retain its importance as an outline of the technical report and, thus, should not deviate significantly from the text. This is especially true in technical reports without an index.

To keep a table of contents unclassified, "(Classified Heading)" should be inserted in lieu of the actual heading in the table of contents if the heading is classified. A "(U)" should appear after the word "Contents." There is no need to place a "(U)" after each head-

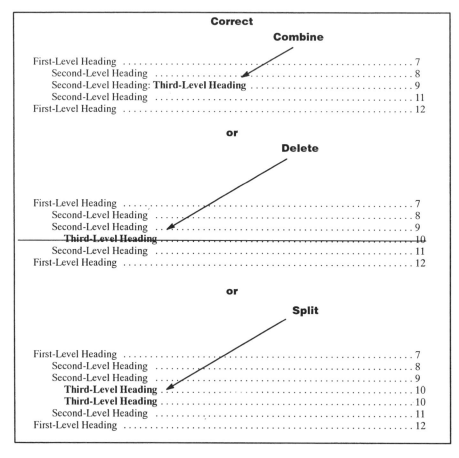

**Figure 3–15.   Minimum Number of Headings**

ing in the table of contents; the omission of a classification marking after each heading indicates that the table of contents is unclassified. If more than a few of the headings are classified, the classified headings should be spelled out and their classification indicated in parentheses. In technical reports using a typographical progression format, the classification of the headings should appear after the headings. In technical reports using a decimal numbering format, the classification of the headings should appear after the section number or appendix letter but before the headings. For unnumbered headings (such as "Abstract" and "Executive Summary") in the same technical reports, the classifications of the headings should appear after the heading. A "(U)" should appear after the word "Contents."

The table of contents in a classified technical report should be consistent with the lists of figures and tables: If the table of contents uses an unclassified format (Figures 3-17 and 3-20), the lists of figures and tables should use an unclassified format; conversely, if the table of contents uses a classified format (Figures 3-18 and 3-21), the lists of figures and tables should use a classified format.

Figure 3-16 shows an example of a table of contents in an unclassified technical report where a typographical progression format is used. Figures 3-17 and 3-18 show, respectively, examples of an unclassified and classified table of contents in a classified technical report using the same type of format.

Figure 3-19 shows an example of a table of contents in an unclassified technical report where a decimal numbering format is used. Figures 3-20 and 3-21 show, respectively, examples of an unclassified and classified table of contents in a classified technical report using the same type of format.

The table of contents should begin on page vii of the front matter.

## 3.18    List(s) of Figures and Tables

ANSI/NISO Z39.18-2005 states that "if a [technical] report contains more than five figures or tables, or some combination totaling more than five, a list of figures and/or tables is required. If a [technical] report contains fewer than five figures or tables, a list is optional. . . . A list of figures precedes a list of tables."[2]

Usually, a defense-related technical report that contains figures and/or tables will include a list of figures and/or tables, regardless of the number. Each list is separate; however, they may placed on the same page if one or both lists are short.

There are similarities between the table of contents and the list(s) of figure and tables:

- The list of figures and list of tables are titled, simply, "Figures" and "Tables" in the technical report.
- Figure and table titles should be different, i.e., the same title should not be used for another figure or table or between figures and tables;

*Contents*

| | *Page* |
|---|---|
| Abstract | v |
| Figures | ix |
| Tables | xi |
| Foreword | xiii |
| Preface/Administrative Information | xv |
| Acknowledgments | xvii |
| Executive Summary | 1 |
|    Second-Level Heading | 1 |
|    Second-Level Heading | 2 |
| Introduction | 3 |
|    Second-Level Heading | 4 |
|    Second-Level Heading | 5 |
|    Second-Level Heading | 6 |
| Methods, Assumptions, and Procedures | 7 |
|    Second-Level Heading | 8 |
|    Second-Level Heading | 9 |
|       Third-Level Heading | 9 |
|       Third-Level Heading | 10 |
|       Third-Level Heading | 10 |
|    Second-Level Heading | 11 |
|    Second-Level Heading | 11 |
| Results and Discussion | 12 |
|    Second-Level Heading | 13 |
|       Third-Level Heading | 13 |
|       Third-Level Heading | 14 |
|    Second-Level Heading | 14 |
|       Third-Level Heading | 15 |
|       Third-Level Heading | 15 |
| Conclusions | 16 |
|    Second-Level Heading | 17 |
|    Second-Level Heading | 17 |
| Recommendations | 18 |
|    Second-Level Heading | 19 |
|    Second-Level Heading | 19 |
| References | 21 |
| Appendix A: Title | 23 |
| Appendix B: Title | 25 |
| Appendix C: Title | 27 |
| Bibliography | 29 |
| Symbols, Abbreviations, and Acronyms | 31 |
| Glossary | 33 |
| Index | 35 |
| Primary Distribution | 37 |

**Figure 3–16.**    Example of Table of Contents in
Unclassified Technical Report Using Typographical
Progression Format

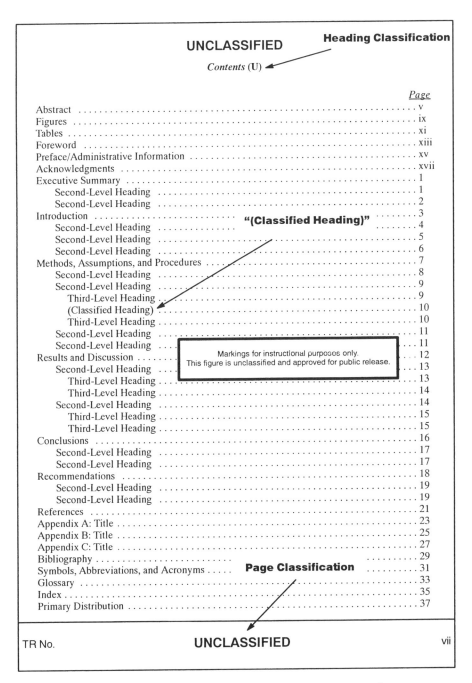

UNCLASSIFIED                    **Heading Classification**

*Contents* (U)

|  | *Page* |
|---|---|
| Abstract | v |
| Figures | ix |
| Tables | xi |
| Foreword | xiii |
| Preface/Administrative Information | xv |
| Acknowledgments | xvii |
| Executive Summary | 1 |
| Second-Level Heading | 1 |
| Second-Level Heading | 2 |
| Introduction | 3 |
| Second-Level Heading | 4 |
| Second-Level Heading | 5 |
| Second-Level Heading | 6 |
| Methods, Assumptions, and Procedures | 7 |
| Second-Level Heading | 8 |
| Second-Level Heading | 9 |
| Third-Level Heading | 9 |
| (Classified Heading) | 10 |
| Third-Level Heading | 10 |
| Second-Level Heading | 11 |
| Second-Level Heading | 11 |
| Results and Discussion | 12 |
| Second-Level Heading | 13 |
| Third-Level Heading | 13 |
| Third-Level Heading | 14 |
| Second-Level Heading | 14 |
| Third-Level Heading | 15 |
| Third-Level Heading | 15 |
| Conclusions | 16 |
| Second-Level Heading | 17 |
| Second-Level Heading | 17 |
| Recommendations | 18 |
| Second-Level Heading | 19 |
| Second-Level Heading | 19 |
| References | 21 |
| Appendix A: Title | 23 |
| Appendix B: Title | 25 |
| Appendix C: Title | 27 |
| Bibliography | 29 |
| Symbols, Abbreviations, and Acronyms | 31 |
| Glossary | 33 |
| Index | 35 |
| Primary Distribution | 37 |

**"(Classified Heading)"**

Markings for instructional purposes only.
This figure is unclassified and approved for public release.

**Page Classification**

TR No.                    UNCLASSIFIED                    vii

**Figure 3-17.**  Example of Unclassified Table of
Contents in Classified Technical Report Using
Typographical Progression Format

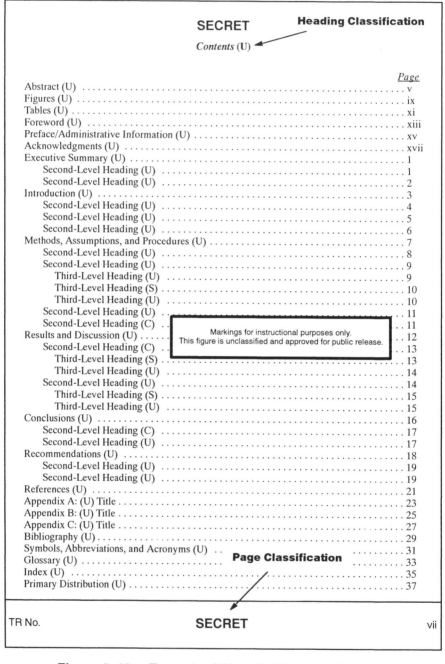

SECRET       **Heading Classification**

*Contents* (U)

|  |  | *Page* |
|---|---|---|
| Abstract (U) | | v |
| Figures (U) | | ix |
| Tables (U) | | xi |
| Foreword (U) | | xiii |
| Preface/Administrative Information (U) | | xv |
| Acknowledgments (U) | | xvii |
| Executive Summary (U) | | 1 |
|   Second-Level Heading (U) | | 1 |
|   Second-Level Heading (U) | | 2 |
| Introduction (U) | | 3 |
|   Second-Level Heading (U) | | 4 |
|   Second-Level Heading (U) | | 5 |
|   Second-Level Heading (U) | | 6 |
| Methods, Assumptions, and Procedures (U) | | 7 |
|   Second-Level Heading (U) | | 8 |
|   Second-Level Heading (U) | | 9 |
|     Third-Level Heading (U) | | 9 |
|     Third-Level Heading (S) | | 10 |
|     Third-Level Heading (U) | | 10 |
|   Second-Level Heading (U) | | 11 |
|   Second-Level Heading (C) | | 11 |
| Results and Discussion (U) | | 12 |
|   Second-Level Heading (C) | | 13 |
|     Third-Level Heading (S) | | 13 |
|     Third-Level Heading (U) | | 14 |
|   Second-Level Heading (U) | | 14 |
|     Third-Level Heading (S) | | 15 |
|     Third-Level Heading (U) | | 15 |
| Conclusions (U) | | 16 |
|   Second-Level Heading (C) | | 17 |
|   Second-Level Heading (U) | | 17 |
| Recommendations (U) | | 18 |
|   Second-Level Heading (U) | | 19 |
|   Second-Level Heading (U) | | 19 |
| References (U) | | 21 |
| Appendix A: (U) Title | | 23 |
| Appendix B: (U) Title | | 25 |
| Appendix C: (U) Title | | 27 |
| Bibliography (U) | | 29 |
| Symbols, Abbreviations, and Acronyms (U) | | 31 |
| Glossary (U) | | 33 |
| Index (U) | | 35 |
| Primary Distribution (U) | | 37 |

> Markings for instructional purposes only.
> This figure is unclassified and approved for public release.

**Page Classification**

TR No.       **SECRET**       vii

**Figure 3–18.** Example of Classified Table of Contents
in Classified Technical Report Using Typographical
Progression Format

## Contents

                                                                          Page
Abstract ........................................................ v
Figures ......................................................... ix
Tables .......................................................... xi
Foreword ....................................................... xiii
Preface/Administrative Information ............................. xv
Acknowledgments ............................................... xvii
Executive Summary ............................................. ES-1
    ES.1   Second-Level Heading ............................... ES-1
    ES.2   Second-Level Heading ............................... ES-2
1   Introduction ............................................. 1-1
    1.1    Second-Level Heading ............................... 1-2
    1.2    Second-Level Heading ............................... 1-3
    1.3    Second-Level Heading ............................... 1-4
2   Methods, Assumptions, and Procedures ..................... 2-1
    2.1    Second-Level Heading ............................... 2-2
    2.2    Second-Level Heading ............................... 2-3
        2.2.1   Third-Level Heading ....................... 2-4
        2.2.2   Third-Level Heading ....................... 2-5
        2.2.3   Third-Level Heading ....................... 2-6
    2.3    Second-Level Heading ............................... 2-7
    2.4    Second-Level Heading ............................... 2-8
3   Results and Discussion ................................... 3-1
    3.1    Second-Level Heading ............................... 3-2
        3.1.1   Third-Level Heading ....................... 3-3
        3.1.2   Third-Level Heading ....................... 3-4
    3.2    Second-Level Heading ............................... 3-5
        3.2.1   Third-Level Heading ....................... 3-6
        3.2.2   Third-Level Heading ....................... 3-7
4   Conclusions .............................................. 4-1
    4.1    Second-Level Heading ............................... 4-2
    4.2    Second-Level Heading ............................... 4-3
5   Recommendations .......................................... 5-1
    5.1    Second-Level Heading ............................... 5-2
    5.2    Second-Level Heading ............................... 5-3
6   References ............................................... 6-1
Appendix A: Title ............................................. A-1
Appendix B: Title ............................................. B-1
Appendix C: Title ............................................. C-1
Bibliography .................................................. Bibliography-1
Symbols, Abbreviations, and Acronyms .......................... SAA-1
Glossary ...................................................... Glossary-1
Index ......................................................... Index-1
Primary Distribution .......................................... PD-1

**Figure 3-19.**   Example of Table of Contents in
Unclassified Technical Report Using Decimal
Numbering Format

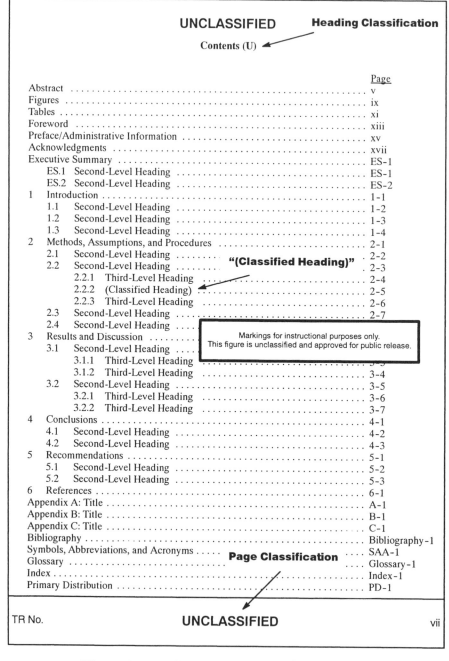

UNCLASSIFIED          **Heading Classification**

Contents (U)

|  | Page |
| --- | --- |
| Abstract | v |
| Figures | ix |
| Tables | xi |
| Foreword | xiii |
| Preface/Administrative Information | xv |
| Acknowledgments | xvii |
| Executive Summary | ES-1 |
|   ES.1   Second-Level Heading | ES-1 |
|   ES.2   Second-Level Heading | ES-2 |
| 1   Introduction | 1-1 |
|   1.1   Second-Level Heading | 1-2 |
|   1.2   Second-Level Heading | 1-3 |
|   1.3   Second-Level Heading | 1-4 |
| 2   Methods, Assumptions, and Procedures | 2-1 |
|   2.1   Second-Level Heading | 2-2 |
|   2.2   Second-Level Heading **"(Classified Heading)"** | 2-3 |
|     2.2.1   Third-Level Heading | 2-4 |
|     2.2.2   (Classified Heading) | 2-5 |
|     2.2.3   Third-Level Heading | 2-6 |
|   2.3   Second-Level Heading | 2-7 |
|   2.4   Second-Level Heading | |
| 3   Results and Discussion | |
|   3.1   Second-Level Heading | |
|     3.1.1   Third-Level Heading | |
|     3.1.2   Third-Level Heading | 3-4 |
|   3.2   Second-Level Heading | 3-5 |
|     3.2.1   Third-Level Heading | 3-6 |
|     3.2.2   Third-Level Heading | 3-7 |
| 4   Conclusions | 4-1 |
|   4.1   Second-Level Heading | 4-2 |
|   4.2   Second-Level Heading | 4-3 |
| 5   Recommendations | 5-1 |
|   5.1   Second-Level Heading | 5-2 |
|   5.2   Second-Level Heading | 5-3 |
| 6   References | 6-1 |
| Appendix A: Title | A-1 |
| Appendix B: Title | B-1 |
| Appendix C: Title | C-1 |
| Bibliography | Bibliography-1 |
| Symbols, Abbreviations, and Acronyms **Page Classification** | SAA-1 |
| Glossary | Glossary-1 |
| Index | Index-1 |
| Primary Distribution | PD-1 |

Markings for instructional purposes only.
This figure is unclassified and approved for public release.

TR No.          UNCLASSIFIED          vii

**Figure 3–20.**   Example of Unclassified Table of
Contents in Classified Technical Report Using Decimal
Numbering Format

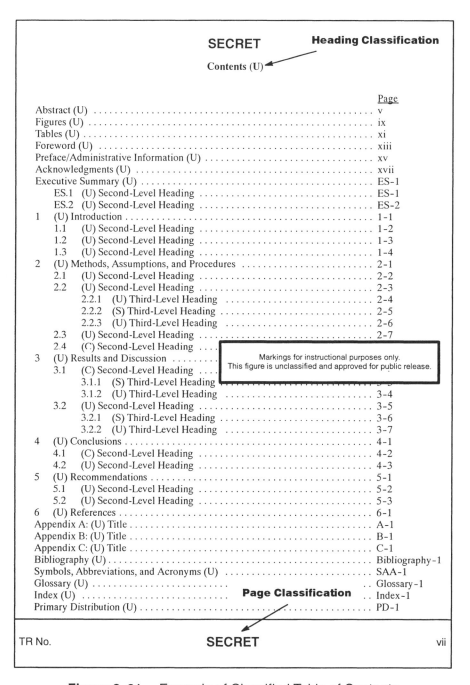

**SECRET**    **Heading Classification**

Contents (U)

| | Page |
|---|---|
| Abstract (U) | v |
| Figures (U) | ix |
| Tables (U) | xi |
| Foreword (U) | xiii |
| Preface/Administrative Information (U) | xv |
| Acknowledgments (U) | xvii |
| Executive Summary (U) | ES-1 |
| ES.1   (U) Second-Level Heading | ES-1 |
| ES.2   (U) Second-Level Heading | ES-2 |
| 1   (U) Introduction | 1-1 |
| 1.1   (U) Second-Level Heading | 1-2 |
| 1.2   (U) Second-Level Heading | 1-3 |
| 1.3   (U) Second-Level Heading | 1-4 |
| 2   (U) Methods, Assumptions, and Procedures | 2-1 |
| 2.1   (U) Second-Level Heading | 2-2 |
| 2.2   (U) Second-Level Heading | 2-3 |
| 2.2.1   (U) Third-Level Heading | 2-4 |
| 2.2.2   (S) Third-Level Heading | 2-5 |
| 2.2.3   (U) Third-Level Heading | 2-6 |
| 2.3   (U) Second-Level Heading | 2-7 |
| 2.4   (C) Second-Level Heading | |
| 3   (U) Results and Discussion | |
| 3.1   (C) Second-Level Heading | |
| 3.1.1   (S) Third-Level Heading | |
| 3.1.2   (U) Third-Level Heading | 3-4 |
| 3.2   (U) Second-Level Heading | 3-5 |
| 3.2.1   (S) Third-Level Heading | 3-6 |
| 3.2.2   (U) Third-Level Heading | 3-7 |
| 4   (U) Conclusions | 4-1 |
| 4.1   (C) Second-Level Heading | 4-2 |
| 4.2   (U) Second-Level Heading | 4-3 |
| 5   (U) Recommendations | 5-1 |
| 5.1   (U) Second-Level Heading | 5-2 |
| 5.2   (U) Second-Level Heading | 5-3 |
| 6   (U) References | 6-1 |
| Appendix A: (U) Title | A-1 |
| Appendix B: (U) Title | B-1 |
| Appendix C: (U) Title | C-1 |
| Bibliography (U) | Bibliography-1 |
| Symbols, Abbreviations, and Acronyms (U) | SAA-1 |
| Glossary (U) | Glossary-1 |
| Index (U) | Index-1 |
| Primary Distribution (U) | PD-1 |

Markings for instructional purposes only.
This figure is unclassified and approved for public release.

**Page Classification**

TR No.    **SECRET**    vii

**Figure 3-21.**   Example of Classified Table of Contents
in Classified Technical Report Using Decimal
Numbering Format

however, subtitles may be identical if they appear in different figures or tables.

- Acronyms should not appear by themselves in a figure or table title, regardless if they have been identified previously in the technical report. Figures and tables are oftentimes used interchangeably among technical reports and, thus, should be considered standalone elements in that context. An acronym, if part of a figure or table title, should be placed in parentheses and preceded by its spelled-out equivalent.
- If the executive summary contains figures and/or tables, these should be included in the list(s) of figures and tables of the technical report. (Figure and/or table titles may be repeated verbatim from the main body of the technical report.)
- If an appendix contains figures and/or tables, these may optionally be included in the list(s) of figures and tables of the technical report (not shown in the examples herein); however, if an appendix contains its own list(s) of figures and tables, e.g., a separate document, its list(s) of figures and tables should not be repeated in the list(s) of figures and tables of the technical report. Also, for consistency in the list(s) of figures and tables of the technical report, if a technical report contains more than one appendix and one of the appendixes contains its own list(s) of figures and tables, then none of the appendix figures and/or tables from the other appendixes should appear in the table of contents of the technical report.
- Periods may be placed after a figure or table number (not shown herein). As in the table of contents, a minimum of three leaders should appear between an entry and its corresponding page number. Runover lines should be indented.
- Subfigure or subtable titles should not be included in the list(s) of figures and tables.
- In classified technical reports, the lists of figures and tables should, if feasible and like the table of contents, be unclassified. To keep the lists of figures and tables unclassified, "(Classified Title )" should be inserted in lieu of the actual title in the lists of figures and tables if the title is classified. A "(U)" should appear after the words "Figures" and "Tables." There is no need to place a "(U)" after each title in the lists of figures and tables; the omission of a classification marking after each title indicates that the lists of figures and tables are unclassified.

  If more than a few of the titles are classified, the classified titles should be spelled out and their classification indicated in parentheses. The classification of the titles should appear after the figure or table numbers but before the titles. A "(U)" should appear after the words "Figures" and "Tables."

The lists of figures and tables in a classified technical report should be consistent with the table of contents and with each other: If the table of contents or one list uses an unclassified format, the other list should use an unclassified format; conversely, if the table of contents or one list uses a classified format, the other list should use a classified format.

In technical reports using a typographical progression format to indicate subordination, figures and tables should be numbered consecutively in the body of the technical report; however, figures and tables in the executive summary should be numbered ES-1, ES-2, etc., as they would in a technical report using a decimal numbering format. In technical reports using a decimal numbering format, figures and tables should be numbered by section. For example, the first figure in Section 1 would be numbered 1-1, and the second figure in Section 1 would be 1-2, etc. The numbering would restart with each section. Tables should be numbered the same way. If there is only one figure or table in one section, it should still be numbered the same way.

Figure 3-22 shows example lists of figures and tables in an unclassified technical report where a typographical progression format is used. Figures 3-23 and 3-24 show, respectively, examples of unclassified and classified lists of figures and tables in a classified technical report using the same type of format.

Figure 3-25 shows example lists of figures and tables in an unclassified technical report where a decimal numbering format is used. Figures 3-26 and 3-27 show, respectively, examples of unclassified and classified lists of figures and tables in a classified technical report using the same type of format.

## 3.19   Foreword

[A] foreword is a conditional introductory statement that presents background material or places in context a [technical] report that is part of a series. It is written by an authority in the field other than the creator [author] of the [technical] report. The name and affiliation of the creator [author] of the foreword follow the last paragraph. A foreword and a preface are not interchangeable, and the information in them is not redundant.[2]

ANSI/NISO Z39.18-2005 states that a foreword is conditional and should be included "when background and context is needed."[2] However, a

*Figures*

| | | Page |
|---|---|---|
| ES-1 | Title | 1 |
| ES-2 | Title | 2 |
| 1 | Title | 3 |
| 2 | Title | 5 |
| 3 | Title | 10 |
| 4 | Title | 15 |
| 5 | Title | 20 |
| 6 | Title | 25 |
| 7 | Title | 30 |

*Tables*

| | | Page |
|---|---|---|
| ES-1 | Title | 1 |
| 1 | Title | 2 |
| 2 | Title | 6 |
| 3 | Title | 11 |
| 4 | Title | 16 |
| 5 | Title | 21 |
| 6 | Title | 26 |
| 7 | Title | 31 |
| 8 | Title | 32 |

TR No.                                                                    ix

**Figure 3-22.**   Example of Lists of Figures and Tables
in Unclassified Technical Report Using Typographical
Progression Format

foreword rarely, if ever, appears in a defense-related technical report and, thus, may be considered optional as opposed to conditional. If a foreword is included, it may provide an opportunity for a supervisor to materially contribute to the technical report rather than being included as a co-author to the technical report when, in fact, the supervisor may have had only minimal input to the technical report and primarily served as a reviewer. A high-

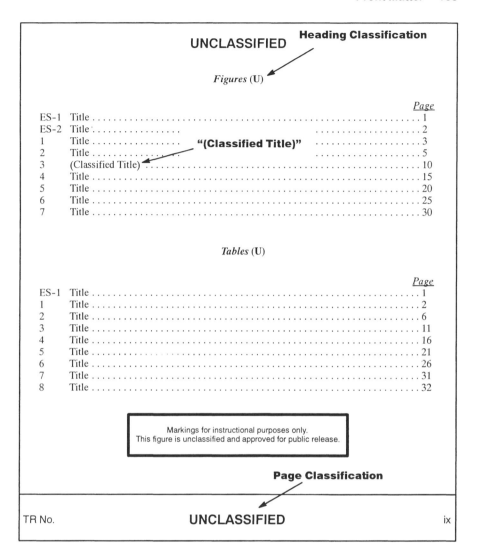

**Figure 3-23.** Example of Unclassified Lists of Figures
and Tables in Classified Technical Report Using
Typographical Progression Format

er-level supervisor may also wish to contribute to a technical report; as such, a technical report may contain more than one foreword.

As ANSI/NISO Z39.18-2005 states, the creator (author) should not write the foreword to the technical report; neither should a co-author. Some organizations routinely require that an author's supervisor be included as a co-author to the technical report. However, if a supervisor writes the foreword, he/she should not be listed as a co-author to the technical report.

SECRET     **Heading Classification**

*Figures* (U)

|        |            | *Page* |
|--------|------------|--------|
| ES-1   | (U) Title  | 1      |
| ES-2   | (U) Title  | 2      |
| 1      | (U) Title  | 3      |
| 2      | (C) Title  | 5      |
| 3      | (U) Title  | 10     |
| 4      | (S) Title  | 15     |
| 5      | (U) Title  | 20     |
| 6      | (U) Title  | 25     |
| 7      | (U) Title  | 30     |

**Title Classification**     *Tables* (U)

|        |            | *Page* |
|--------|------------|--------|
| ES-1   | (U) Title  | 1      |
| 1      | (U) Title  | 2      |
| 2      | (U) Title  | 6      |
| 3      | (S) Title  | 11     |
| 4      | (U) Title  | 16     |
| 5      | (U) Title  | 21     |
| 6      | (C) Title  | 26     |
| 7      | (U) Title  | 31     |
| 8      | (U) Title  | 32     |

> Markings for instructional purposes only.
> This figure is unclassified and approved for public release.

**Page Classification**

TR No.              SECRET              ix

**Figure 3-24.**   Example of Classified Lists of Figures
and Tables in Classified Technical Report Using
Typographical Progression Format

The foreword should be written in the first person and in the present
tense. The foreword may contain unclassified controlled information; however, it should be unclassified in classified technical reports. Figures 3-28
and 3-29 show, respectively, an example of a foreword in an unclassified
and classified technical report.

**Figures**

|       |       | Page  |
|-------|-------|-------|
| ES-1  | Title | ES-1  |
| ES-2  | Title | ES-2  |
| 1-1   | Title | 1-1   |
| 1-2   | Title | 1-2   |
| 2-1   | Title | 2-1   |
| 3-1   | Title | 3-1   |
| 3-2   | Title | 3-2   |
| 4-1   | Title | 4-1   |
| 5-1   | Title | 5-1   |

**Tables**

|       |       | Page  |
|-------|-------|-------|
| ES-1  | Title | ES-1  |
| 1-1   | Title | 1-1   |
| 2-1   | Title | 2-1   |
| 2-2   | Title | 2-2   |
| 2-3   | Title | 2-3   |
| 3-1   | Title | 3-1   |
| 4-1   | Title | 4-1   |
| 4-2   | Title | 4-2   |
| 5-1   | Title | 5-1   |

TR No.                                                              ix

**Figure 3-25.** Example of Lists of Figures and Tables
in Unclassified Technical Report Using Decimal
Numbering Format

## 3.20 Preface (Administrative Information)

A preface is a conditional introductory statement that announces the
purpose and scope of the [technical] report and acknowledges the contributions of individuals not identified as authors/creators or editors.
Sometimes a preface specifies the audience for which a [technical] re-

UNCLASSIFIED    **Heading Classification**

Figures (U)

| | | Page |
|---|---|---|
| ES-1 | Title | ES-1 |
| ES-2 | Title **"(Classified Title)"** | ES-2 |
| 1-1 | Title | 1-1 |
| 1-2 | Title | 1-2 |
| 2-1 | (Classified Title) | 2-1 |
| 3-1 | Title | 3-1 |
| 3-2 | Title | 3-2 |
| 4-1 | Title | 4-1 |
| 5-1 | Title | 5-1 |

Tables (U)

| | | Page |
|---|---|---|
| ES-1 | Title | ES-1 |
| 1-1 | Title | 1-1 |
| 2-1 | Title | 2-1 |
| 2-2 | Title | 2-2 |
| 2-3 | Title | 2-3 |
| 3-1 | Title | 3-1 |
| 4-1 | Title | 4-1 |
| 4-2 | Title | 4-2 |
| 5-1 | Title | 5-1 |

> Markings for instructional purposes only.
> This figure is unclassified and approved for public release.

**Page Classification**

TR No.    UNCLASSIFIED    ix

**Figure 3-26.** Example of Unclassified Lists of Figures and Tables in Classified Technical Report Using Decimal Numbering Format

port is intended; it may also highlight the relationship of the [technical] report to a specific project or program.[2]

Like a foreword, ANSI/NISO Z39.18-2005 states that a preface is conditional and should be included "when background and context is needed."[2] However, some of the administrative information normally included in a preface is essential to a defense-related technical report. Thus, "administra-

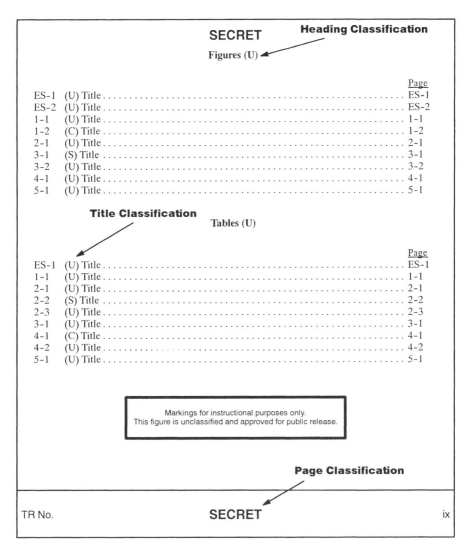

**Figure 3-27.** Example of Classified Lists of Figures
and Tables in Classified Technical Report Using
Decimal Numbering Format

tive information" or a comparable title is occasionally used in lieu of "preface" in defense-related technical reports and is a required element of the technical report. Most of the administrative information can be obtained from item 5 on the report documentation page or Standard Form 298 (Section 3.15). This includes the following numbers:

a.  Contract number

b.  Grant number

   c.  Program element number

   d.  Project number

   e.  Task number

   f.  Work unit number.

In addition, the name and address (city, State, and ZIP code) of the DoD sponsor, if different from the performing organization, should be included in the preface. Because of the frequency of reassignments within DoD, names of individuals should not be included in DoD sponsor information; rather, the code of the individual followed by the name and address of the sponsoring organization are all that is required. The status of the technical report, itself, e.g., "interim," "final," etc., should also be indicated in the preface.

Because of its purpose and need to include administrative information, the preface in defense-related technical reports usually does not additionally acknowledge the contributions of individuals "not identified as authors/creators or editors,"[2] regardless of length; this should be done in "Acknowledgments," which is explained in Section 3.21.

The preface should be written by the author in the third person and in the past tense. The preface may contain unclassified controlled information; however, it should be unclassified in classified technical reports. Figures 3–28 and 3–29 show, respectively, an example of a preface in an unclassified and classified technical report.

## 3.21   Acknowledgments

ANSI/NISO Z39.18–2005 states that "acknowledgments of technical assistance that contributed to the content of the [technical] report are made at an appropriate place in the preface or in the text; however, lengthy acknowledgments are often made in a conditional section titled 'Acknowledgments.'"[2] ANSI/NISO Z39.18–2005 adds that acknowledgments are conditional and should be included "when significant."[2]

Notwithstanding ANSI/NISO Z39.18–2005 and as implied previously, acknowledgments should not appear in the body of defense-related technical reports. Acknowledgments seldom appear in defense-related technical reports and may be considered optional as opposed to conditional; however, acknowledgments appear more often than a foreword.

Acknowledgments should be brief. Illustrators, like editors, and others who assisted in the preparation of the technical report as part of their normal work routine should not be acknowledged. Organizations or groups may be acknowledged in lieu of individuals; however, the acknowledgement should be specific. If included, acknowledgments should be the last element of the front matter.

The acknowledgments should be written in the third person and in the present tense. Third-person phrases such as "the author wishes to thank" are awkward and should be avoided; instead, use of the passive voice may be necessary. The acknowledgments may contain unclassified controlled information; however, the acknowledgments should be unclassified in classified technical reports. Figures 3-28 and 3-29 show, respectively, an example of acknowledgments in an unclassified and classified technical report.

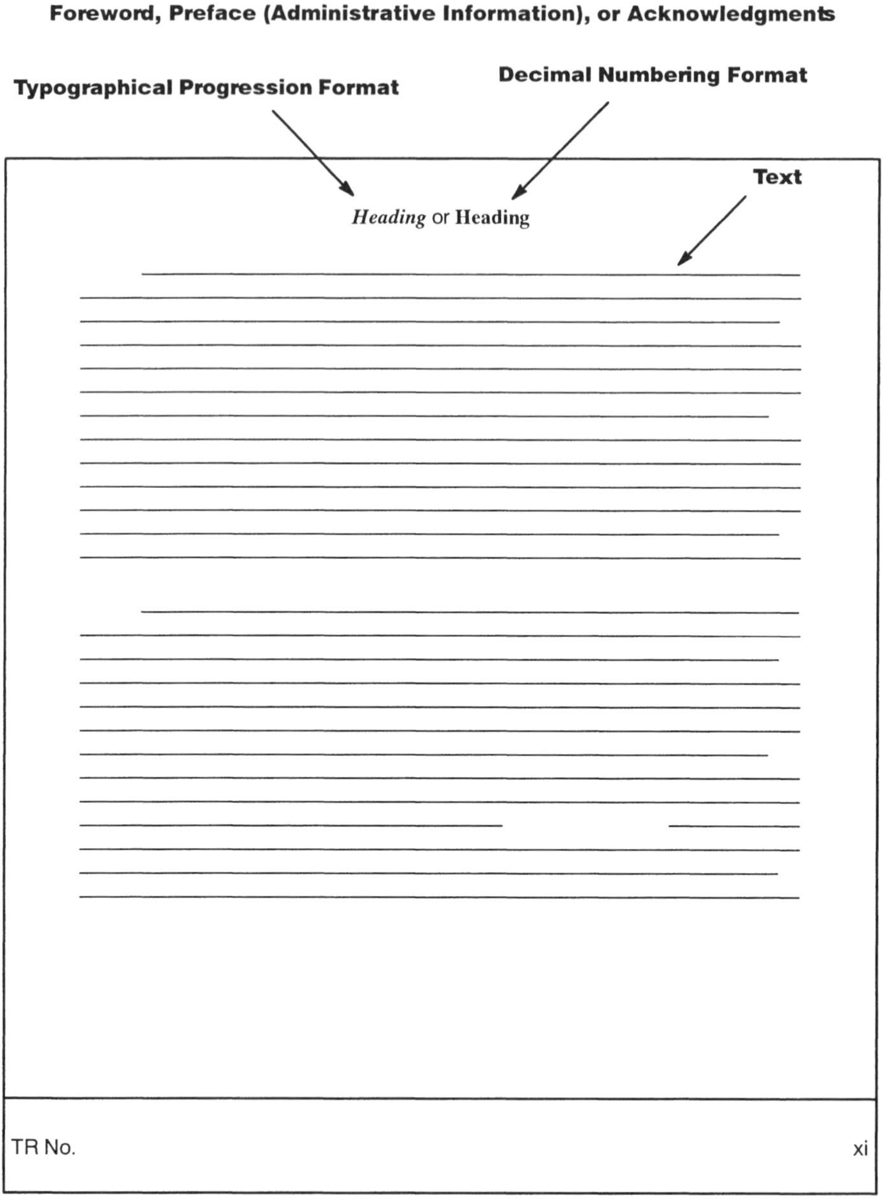

**Foreword, Preface (Administrative Information), or Acknowledgments**

**Typographical Progression Format**

**Decimal Numbering Format**

**Text**

*Heading* or Heading

TR No.                                                                                           xi

**Figure 3-28.**   Example of Foreword, Preface
(Administrative Information), or Acknowledgments in
Unclassified Technical Report

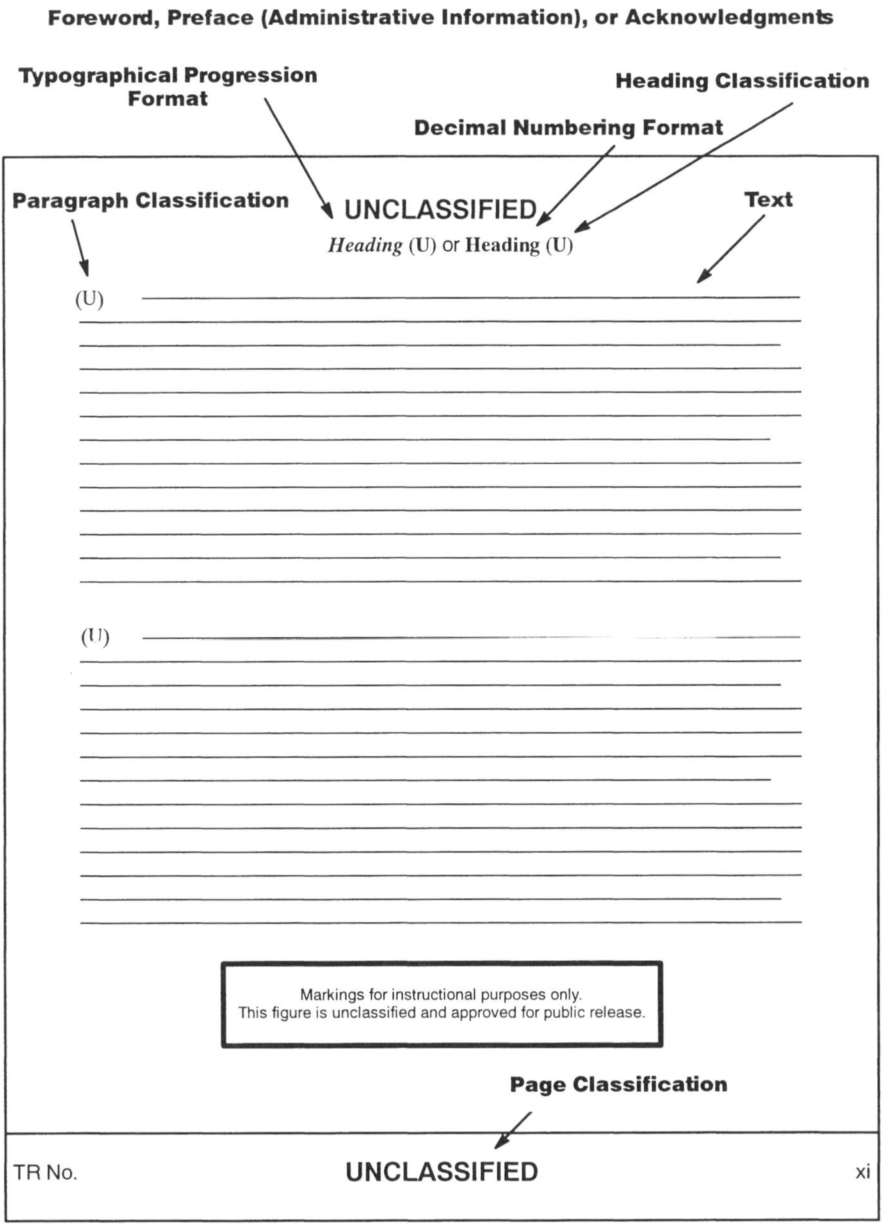

**Foreword, Preface (Administrative Information), or Acknowledgments**

**Typographical Progression
Format**

**Heading Classification**

**Decimal Numbering Format**

**Paragraph Classification**

UNCLASSIFIED

**Text**

*Heading* (U) or Heading (U)

(U) _____

_____

_____

_____

_____

_____

_____

_____

_____

_____

_____

_____

(U) _____

_____

_____

_____

_____

_____

_____

_____

_____

_____

_____

Markings for instructional purposes only.
This figure is unclassified and approved for public release.

**Page Classification**

TR No.                    UNCLASSIFIED                    xi

**Figure 3-29.** Example of Foreword, Preface
(Administrative Information), or Acknowledgments in
Classified Technical Report

# Chapter *4*

# *Body*

The body is the part of the [technical] report in which the creator [author] describes methods, assumptions, and procedures, then presents and discusses the results and draws conclusions and recommends actions based on those results. The organization of a [technical] report depends on its subject matter and audience as well as its purpose. . . . Thus, the organization of the content may vary widely and the organization of the [technical] report may be divided into sections or chapters.[2]

Notwithstanding ANSI/NISO Z39.18-2005, the organization of a defense-related technical report should follow a certain logic and purpose, and arbitrary deviations in organization should be avoided. The body of defense-related technical reports should contain the following elements in the order presented:

a. Executive summary
b. Introduction
c. Methods, assumptions, and procedures
d. Results and discussion
e. Conclusions
f. Recommendations
g. List of references (conditional).

Elements b to f may not appear verbatim, e.g., "Background" may be used in lieu of "Introduction"; "Results and Discussion" may be divided into two elements, "Results" and "Discussion of Results"; and certain elements may be combined into one element, e.g., "Conclusions and Recommendations." However, the order of presentation should remain the same.

## 4.1 Executive Summary

ANSI/NISO Z39.18-2005 refers to a required "summary"[2] and an optional, separate "executive summary"[2] that is "prepared for a management-level audience"[2] in technical reports that exceed 50 pages. ANSI/NISO Z39.18-2005 describes a summary as follows:

> The summary differs from the abstract in purpose, audience, and length. Because the summary restates key points, material not included in the text [body] does not appear in the summary. Introductory material (purpose, scope, and organization), descriptive material (nature and method of investigation), and the most important results and conclusions are summarized, with emphasis on the findings of the research and recommendations.[2]

ANSI/NISO Z39.18-1995 adds that "the length of the summary typically does not exceed 2 percent of the body of the [technical] report."[4] This percentage is more applicable to very large technical reports. In shorter technical reports, the percentage may increase proportionally to 10 percent of the body of the technical report.

ANSI/NISO Z39.18-2005 defines a summary and an executive summary: A summary "clearly states the key points of the [technical] report—including the problem under investigation, the principal results and conclusions, and a recommended course of action for decision makers,"[2] while "an Executive Summary is a non-technical presentation that provides an adequate level of detail for decision makers needing a basic understanding of a research problem and the major findings but who do not plan to read the [technical] report in its entirety."[2] Thus, while ANSI/NISO Z39.18-2005 states that a summary and an executive summary are directed to decisionmakers, i.e., executives, it seemingly makes a "technical" versus "non-technical" distinction between the two.

In reality, the difference between a "technical" summary and a "non-technical" executive summary is subtle and probably precludes the need for both. Decisionmakers in the defense community usually have more than a superficial knowledge of a subject under their purview and should not require a "non-technical" presentation. Notwithstanding, defense-related technical reports may contain a summary and an executive summary; however, the inclusion of both is extremely rare, if at all, regardless of the number of pages in the technical report. As defined herein, a "summary" and "executive summary" are synonymous; the more popular and prevalent term—executive summary—is used herein and should be used to describe this element of a defense-related technical report if it contains only a single summary.

Unlike an abstract, figures and tables may be included in an executive summary; however, like an abstract, an executive summary is a standalone element in a technical report and, like an abstract, should not provide brief descriptions of the elements of the main body of the technical report, e.g., "Section 1 provides an introduction." ANSI/NISO Z39.18-2005 reinforces the standalone aspect of the executive summary:

> Although a summary depends on the content in that it introduces no new information, it is independent from the user's point of view; therefore, all symbols, abbreviations, and acronyms are defined, and unusual terms are explained. A summary does not contain references or cross-references to other [elements] of the [technical] report.[2]

While it is preferable that an abstract be unclassified in a classified technical report, an executive summary may be classified in whole or in part. Also, an executive summary may contain unclassified controlled information.

The executive summary should be written after the rest of the technical report but before the abstract. Like the abstract, the executive summary should be written in the third person and in the present and past tense. Those portions of the executive summary that refer to the introduction and the recommendations elements in the technical report should be written in the present tense; the rest of the executive summary should be written in the past tense.

In technical reports using a typographical progression format, the executive summary begins on a new odd-numbered, right-hand page and is paginated 1, 2, etc. In technical reports using a decimal numbering format, the executive summary also begins on a new odd-numbered, right-hand page but is paginated ES-1, ES-2, etc., to distinguish it from the main body of the technical report. In the decimal numbering format, the executive summary is not considered Section 1 but, rather, is simply an unnumbered section.

Figures 4-1 and 4-2 show, respectively, examples of an executive summary in an unclassified and classified technical report using a typographical progression format. Figures 4-3 and 4-4 show, respectively, examples of an executive summary in an unclassified and classified technical report using a decimal numbering format.

## 4.2    Introduction

The required introduction provides readers with general information they need to understand more detailed information in the rest of the [technical] report. It introduces the subject, purpose, scope, and the

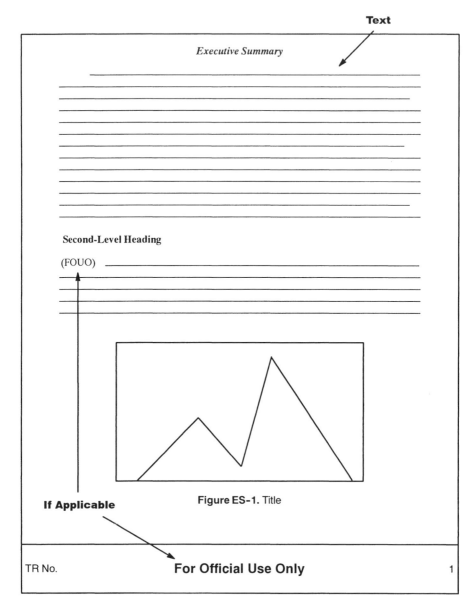

**Figure 4-1.**  Example of Executive Summary in
Unclassified Technical Report Using Typographical
Progression Format

Second-Level Heading

Table ES-1. Title

2

TR No.

**Figure 4-1.** Example of Executive Summary in Unclassified Technical Report Using Typographical Progression Format (Continued)

way the author/creator plans to develop the topic. The introduction also indicates the audience for the [technical] report: who is expected to read it and act on its recommendations or review its findings. . . . The introduction does not, however, include findings, conclusions, or recommendations.[2]

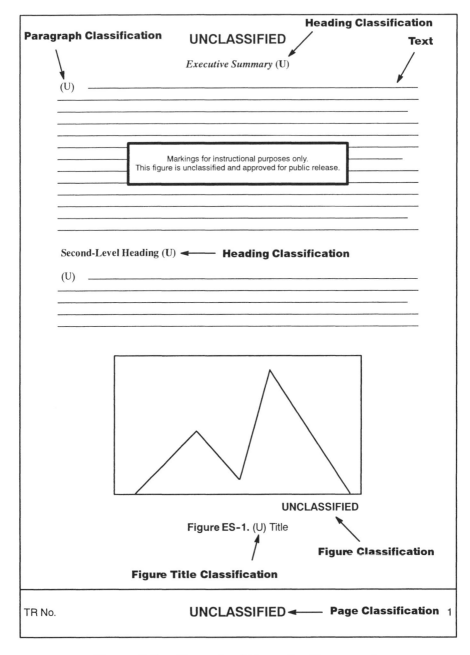

**Figure 4-2.**   Example of Executive Summary in
Classified Technical Report Using Typographical
Progression Format

UNCLASSIFIED ◄——— **Page Classification**

**Second-Level Heading (U)**

(U) _____

Markings for instructional purposes only.
This figure is unclassified and approved for public release.

(U) _____

**Table Title Classification**

**Table ES-1.** (U) Title

\*

†

\*
†

UNCLASSIFIED

**Table Classification**

2    **UNCLASSIFIED**    TR No.

**Figure 4-2.**   Example of Executive Summary in
Classified Technical Report Using Typographical
Progression Format (Continued)

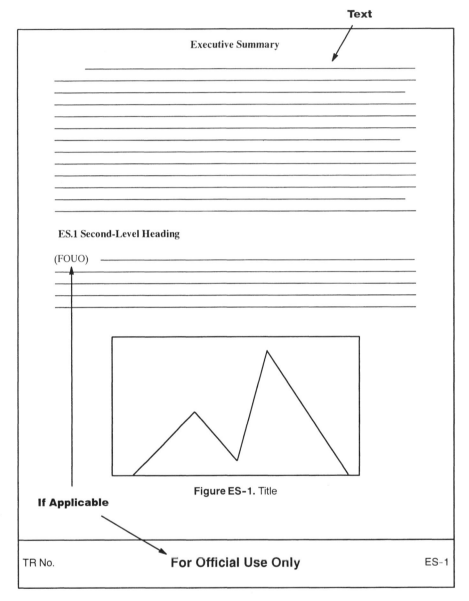

**Figure 4-3.**   Example of Executive Summary in
Unclassified Technical Report Using Decimal
Numbering Format

The last sentence is noteworthy because some authors are mistakenly inclined to include their findings, conclusions, or recommendations in the

**ES.2 Second-Level Heading**

**Table ES-1.** Title

Figure 4-3.  Example of Executive Summary in
Unclassified Technical Report Using Decimal
Numbering Format (Continued)

ES-2                                                    TR No.

introduction—basically equating the introduction with the abstract and
executive summary.

ANSI/NISO Z39.18-2005 adds the following: "The statement of the subject defines the topic and associated terminology and may include the
theory behind the subject, its historical background, its significance, and a
review of pertinent literature."[2] The fourth and last phrase was added to the

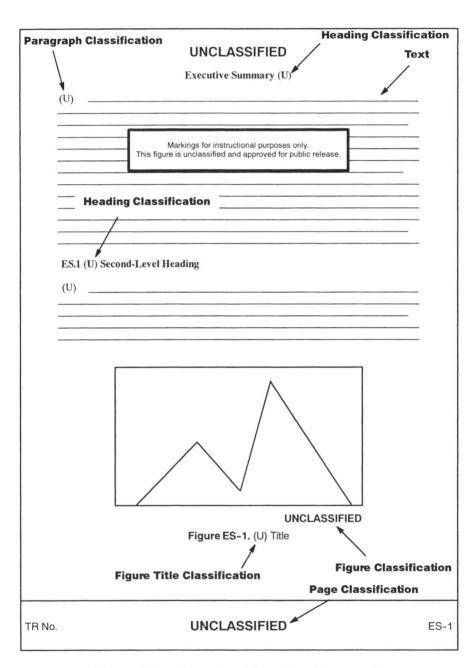

**Figure 4-4.** Example of Executive Summary in
Classified Technical Report Using Decimal Numbering
Format

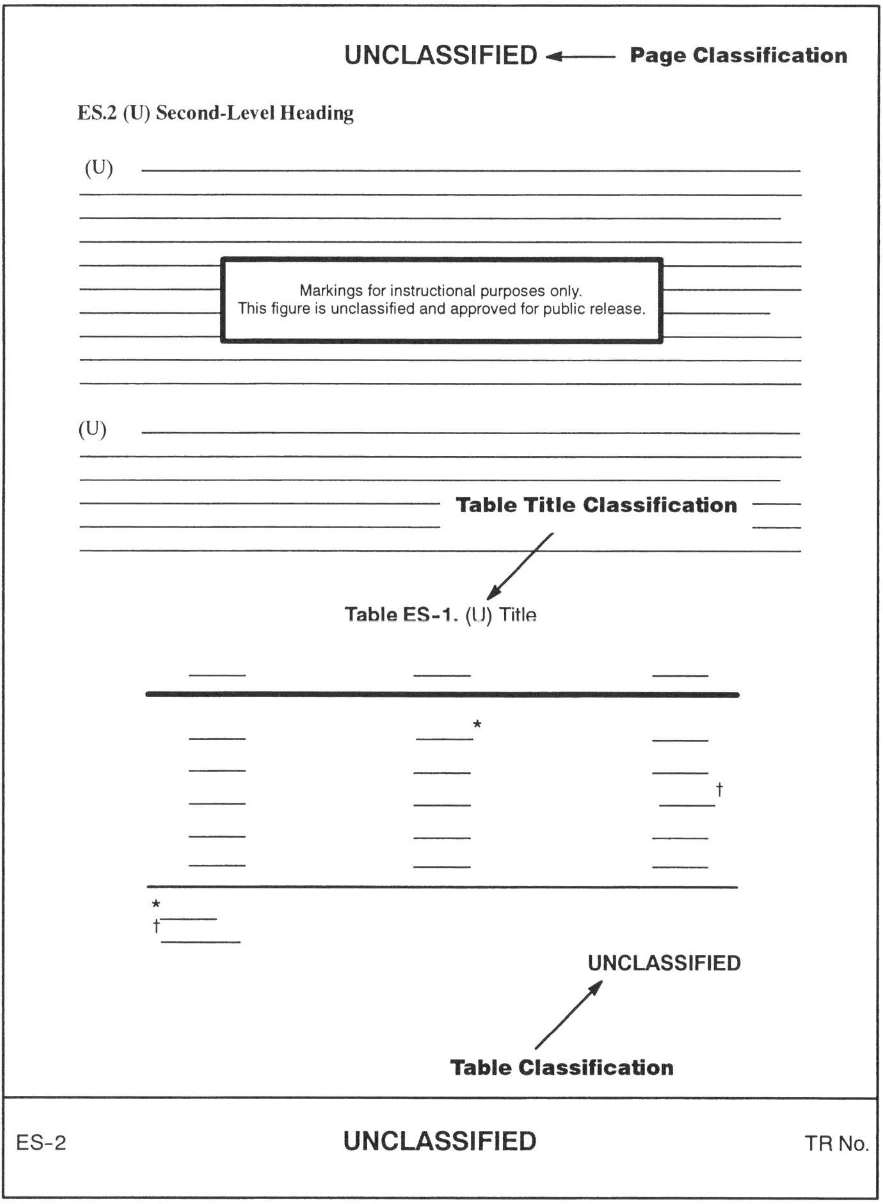

**Figure 4-4.**   Example of Executive Summary in
Classified Technical Report Using Decimal Numbering
Format (Continued)

standard when it was revised in 2005 and is significant. In writing the introduction, the author should perform an exhaustive literature search or review of the subject that includes DTIC's resources and in-house libraries or TICs and by perusing known earlier related work performed by colleagues and by the author. As a result, most, if not all, references should appear in the introduction.

In technical reports using a typographical progression format, the introduction begins on an odd-numbered, right-hand page but inherits or continues the page numbers from the executive summary. Therefore, if the executive summary ends on page 3, the introduction should begin on page 5. The introduction constitutes Section 1 in technical reports using a decimal numbering format. It begins on an odd-numbered, right-hand page and is numbered 1-1, 1-2, etc.

The introduction should be written in the third person and in the past and present tense. When citing previous work, the work, itself, should be discussed in the past tense and published, accepted results should be discussed in the present tense, e.g., "Einstein theorized [past tense] that energy equals [present tense] mass times the speed of light squared." The rest of the introduction should be written in the past tense.

Where applicable, an author and single co-author should be cited by surname. The term "et al." may be substituted for the surnames of two or more co-authors. However, if possible, "et al." should not appear in the list of references (Section 4.8); instead, all co-authors should be identified.

## 4.3   Methods, Assumptions, and Procedures

A description of the methods, assumptions, and procedures used in an investigation is a required [element]. A succinct explanation of them enables readers to evaluate the results without referring extensively to the references. The description should be complete enough that a knowledgeable reader could duplicate the procedures of the investigation.[2]

The methods, assumptions, and procedures are the core of the technical report. The ability to "duplicate the procedures of the investigation"[2] is the basis for a technical report.

The system of measurement (for example, metric or English [standard]) is identified. If the research included apparatus, instruments, or reagents, a description of the apparatus, the design and precision of the instruments, and the nature of the reagents are explained in this required section [element] of the text [body].[2]

Some technical reports include the metric equivalent for an English or standard unit of measurement and vice versa, usually in parentheses. However, while the trend toward universal adoption of metric units of measurement continues, the need to include both units of measurement in a technical report is oftentimes unnecessary and may be confusing, especially when the two units of measurement are not exactly equivalent, e.g., degrees Fahrenheit and degrees Celsius.

Proprietary terms or trade names should, if possible, be avoided in a defense-related technical report. They are most likely to appear in the methods, assumptions, and procedures. Instead, non-proprietary, generic terms should be used. Furthermore, some defense-related organizations prohibit the unnecessary mention of proprietary terms or trade names in their technical reports unless certain related products are being tested and evaluated, which should be indicated in the distribution statement.

The methods, assumptions, and procedures should be written in the third person and in the past tense.

## 4.4    Results and Discussion

As stated previously, this element may be divided into two elements or presented as one element.

> The results section [element] presents the findings based on the methods. The discussion section [element] indicates the degree of accuracy and the significance of the results of the research described. Specific values used to substantiate conclusions appear in the body [this element]. Supporting details not essential to an understanding of the results appear in an appendix. . . . The discussion accounts for the results but does not interpret them.[2]

If the materials, methods, and assumptions are the core of a defense-related technical report, the results and discussion are its most significant element. The conclusions and recommendations are based on the results. The results and discussion element will normally contain the largest number of figures and tables in a defense-related technical report.

The results and discussion should be written in the third person and in the past tense.

## 4.5    Conclusions

> The required conclusions section [element]interprets findings that have been substantiated in the discussion of results and discusses their implications. The section [element] introduces no new material other

than remarks based on these findings. It includes the author's/creator's opinions and is written to be read independently of the text [body]. The section [element] could include a summary of the conclusions from similar studies, a conclusion based solely on the current results, or an overall conclusion.[2]

While the conclusions may be "written to be read independently of the text [body],"[2] they are not a standalone element, such as an abstract or executive summary, and, therefore, may reference other elements in the technical report and "similar studies."[2] Furthermore, the conclusions should not summarize the technical report.

The conclusions should be written in the third person and in the past tense.

## 4.6 Recommendations

The . . . recommendations section [element] presents a course of action based on the results and conclusions of the study. . . . Recommendations might include additional areas for study, alternate design approaches, or production decisions. Specific recommendations are presented in a numbered or bulleted list that is introduced by an informative lead-in sentence. Recommendations may also be included within the conclusion section.[2]

The recommendations are directed to the decisionmakers in an organization and, thus, comprise an essential part of the executive summary. The recommendations should, like the conclusions, not summarize the technical report. Also, all recommendations should be grouped in this element and not appear separately or repeated in other elements of the main body of the technical report. The distinction between numbered or alphanumeric lists and bullet lists is explained in Chapter 2, Section 2.2.1.2.3.

The recommendations should be written in the third person and in the present and, possibly, future tenses. Phrases such as "it is recommended" are vague and awkward and should be avoided; instead, an organization "should" proceed as directed by the author. Occasionally, the recommendations discuss an anticipated course of action by an organization; here, the future tense may be appropriate.

## 4.7 Examples of Main Body

Figures 4-5 and 4-6 show, respectively, condensed examples of the main body (introduction; methods, assumptions, and procedures; results and discussion; conclusions; and recommendations) in an unclassified and classified technical report using a typographical progression format.

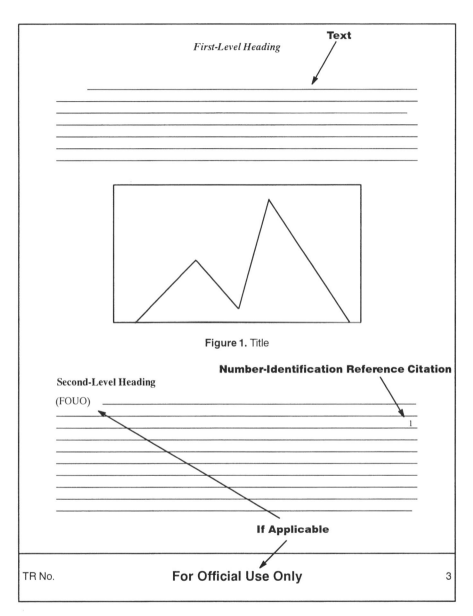

**Figure 4-5.**    Example of Main Body in Unclassified
Technical Report Using Typographical Progression
Format

**Second-Level Heading**

**Bullet Listing**

- 
- 
- 
- 

**Table 1.** Title

4

TR No.

**Figure 4-5.**   Example of Main Body in Unclassified
Technical Report Using Typographical Progression
Format (Continued)

Figures 4-7 and 4-8 show, respectively, condensed examples of the main body in an unclassified and classified technical report using a decimal numbering format.

*Third-Level Heading*

*Fourth-Level Heading.*

*Fourth-Level Heading.*

*Third-Level Heading*

TR No.                                                                              5

**Figure 4-5.**   Example of Main Body in Unclassified
Technical Report Using Typographical Progression
Format (Continued)

## 4.8    List of References

The list of references is a conditional element; however, most defense-related technical reports cite outside work, so a list of references is highly probable. The list of references should contain only those documents cited in the main body of the technical report, exclusive of the executive summa-

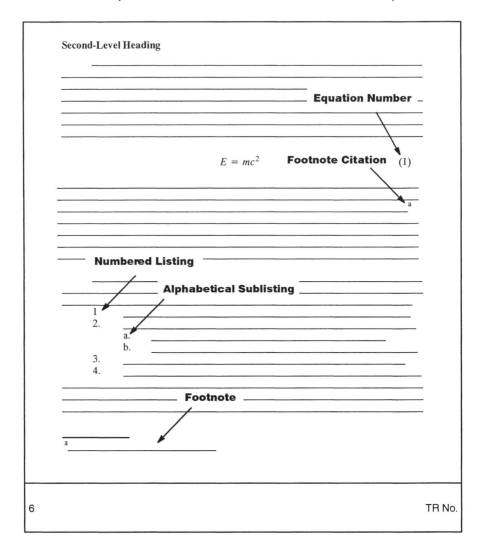

**Figure 4-5.**   Example of Main Body in Unclassified
Technical Report Using Typographical Progression
Format (Continued)

ry. Also, it should not contain documents cited only in an appendix. Documents appearing in the list of references should be readily available to the reader, if not necessarily in the public domain. Therefore, draft documents; personal communications, e.g., e-mail and telephone conversations; and internal documents, e.g., memorandums, should not be included in the list of references. Instead, they should appear as footnotes in the main body of the technical report.

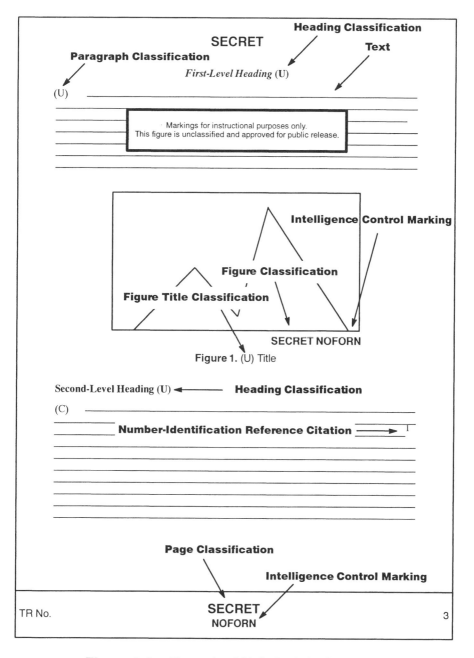

**Figure 4-6.**   Example of Main Body in Classified
Technical Report Using Typographical Progression
Format

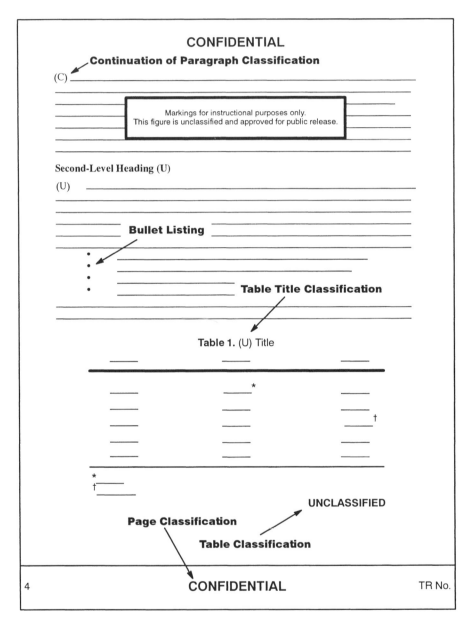

**Figure 4–6.**   Example of Main Body in Classified
Technical Report Using Typographical Progression
Format (Continued)

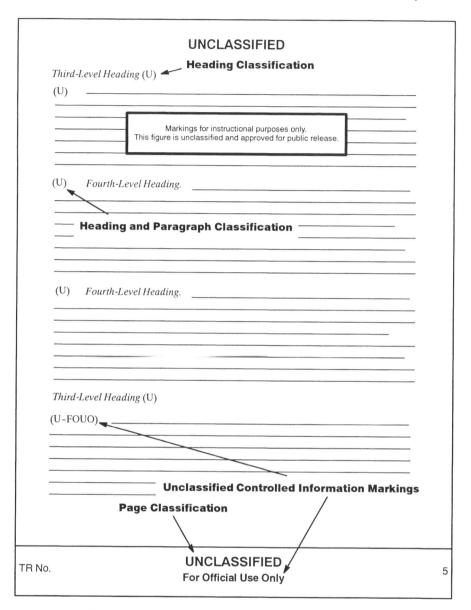

**Figure 4-6.** Example of Main Body in Classified
Technical Report Using Typographical Progression
Format (Continued)

If a technical report contains a list of references, it is the last element of the body of the technical report. It should begin on a new right-hand, odd-numbered page in a technical report using a typographical progression for-

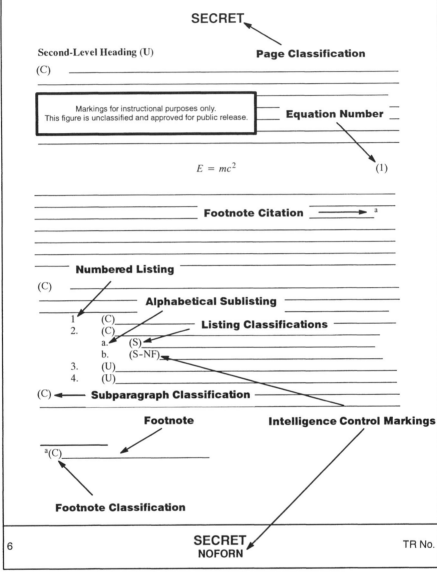

**Figure 4-6.**   Example of Main Body in Classified
Technical Report Using Typographical Progression
Format (Continued)

mat. In a technical report using a decimal numbering format, it should be
the last numbered section.

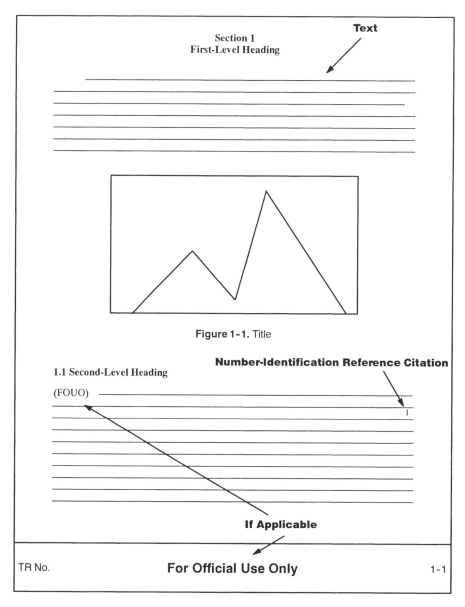

**Figure 4-7.** Example of Main Body in Unclassified
Technical Report Using Decimal Numbering Format

The list of references is not an appendix and should not be identified as such. It is usually titled, simply, "References"; however, it may be called "Sources"[2] or "Works Cited."[2] ANSI/NISO Z39.18–2005 also states that it may be called "Bibliography"[2]; however, as defined herein, while the list of

**1.2 Second-Level Heading**

**Bullet Listing**

Table 1-1. Title

1-2 TR No.

**Figure 4-7.** Example of Main Body in Unclassified
Technical Report Using Decimal Numbering Format
(Continued)

references contains bibliographic information, it is not a bibliography and vice versa. As one of the alternative titles states, it is a list of works or documents cited in the technical report. A bibliography, which is described in Chapter 5, is a supplementary list of works not cited in the technical report; however, a technical report may contain a bibliography and not contain a list of references.

**1.2.1 Third-Level Heading**

_____

_____

_____

_____

_____

**1.2.1.1 Fourth-Level Heading**

_____

_____

_____

_____

_____

**1.2.1.2 Fourth-Level Heading**

_____

_____

_____

_____

_____

**1.2.2 Third-Level Heading**

_____

_____

_____

_____

_____

_____

TR No.                                                                 1–3

**Figure 4-7.** Example of Main Body in Unclassified
Technical Report Using Decimal Numbering Format
(Continued)

ANSI/NISO Z39.18–2005 states that a reference should include "[the] name of [the] author(s)/creator(s), [the] title of [the] referenced work, and publication data or digital-access information."[2] While the author of the technical report should attempt to obtain the referenced information first-hand and include all of the aforementioned information in the reference, there are occasions when the referenced information can only be obtained

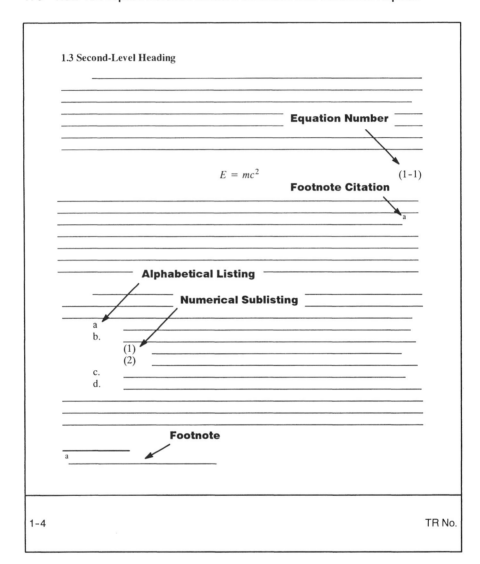

**Figure 4-7.**   Example of Main Body in Unclassified
Technical Report Using Decimal Numbering Format
(Continued)

secondhand and only some of the information is available. For example, some references do not indicate an author per se, while some references do not list co-authors, using "et al." instead. Also, titles oftentimes change from draft to publication or are not copied verbatim; this is most noticeable in secondhand information. Ultimately, the publication data or digital-access information is usually the most accurate information in a reference.

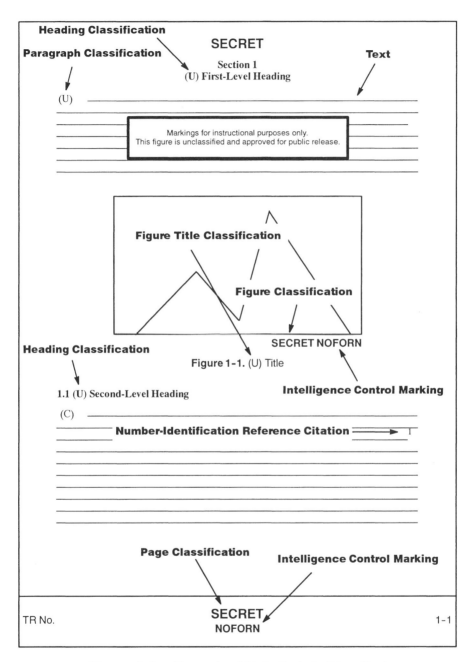

**Figure 4-8.**  Example of Main Body in Classified
Technical Report Using Decimal Numbering Format

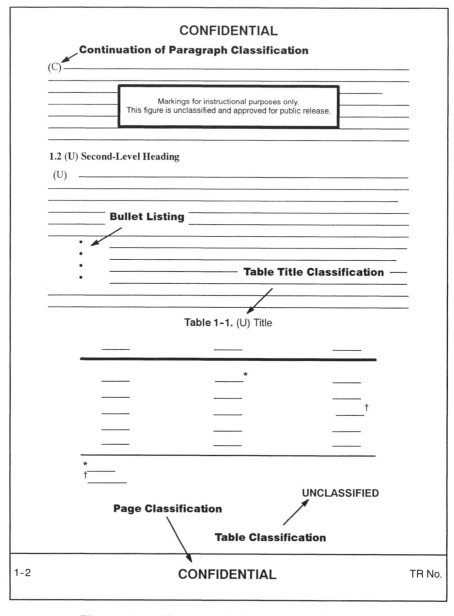

**Figure 4-8.**   Example of Main Body in Classified
Technical Report Using Decimal Numbering Format
(Continued)

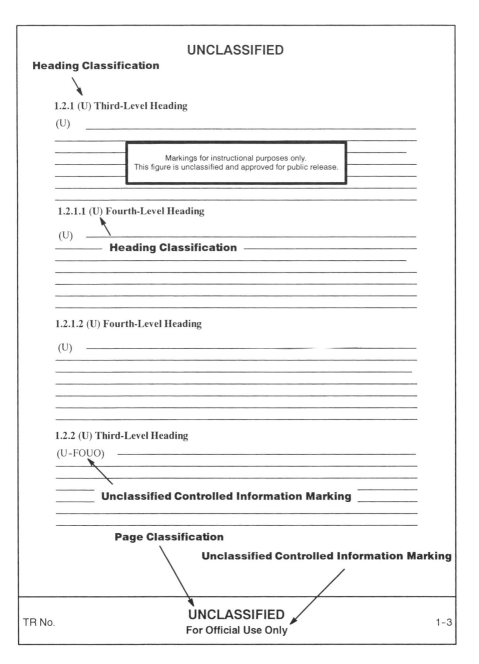

UNCLASSIFIED

**Heading Classification**

1.2.1 (U) Third-Level Heading

(U)

Markings for instructional purposes only.
This figure is unclassified and approved for public release.

1.2.1.1 (U) Fourth-Level Heading

(U)

**Heading Classification**

1.2.1.2 (U) Fourth-Level Heading

(U)

1.2.2 (U) Third-Level Heading

(U-FOUO)

**Unclassified Controlled Information Marking**

**Page Classification**

**Unclassified Controlled Information Marking**

TR No.

UNCLASSIFIED
**For Official Use Only**

1-3

**Figure 4-8.**   Example of Main Body in Classified
Technical Report Using Decimal Numbering Format
(Continued)

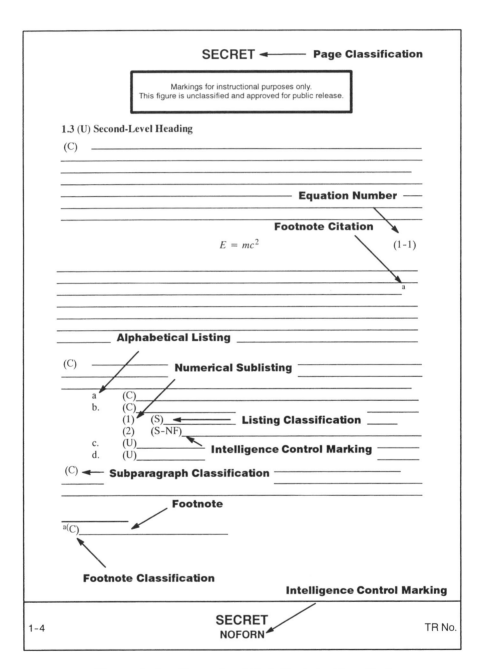

**Figure 4–8.** Example of Main Body in Classified
Technical Report Using Decimal Numbering Format
(Continued)

Basically, enough accurate information should be provided in a reference so that the reader can obtain the referenced work without performing an exhaustive search-and-retrieval effort.

If the referenced work is a Government document approved for public release, the National Technical Information Service (NTIS) number should be included, when available.[2] Also, if the referenced work is included in DTIC's technical report database, the accessioned document number assigned by DTIC should be included.

ANSI/NISO Z39.18–2005 offers three forms of referencing: the number-identification system, the author-date format, and a footnote, endnote, or referenced link used in conjunction with a bibliography. While the latter two formats are acceptable for many types of technical reports, the first—the number-identification system—is best suited for defense-related technical reports for the following reasons:

- The number-identification system is a simple, compact system whereby each reference is identified—usually by a superscript Arabic number—in the main body of the technical report, and a corresponding list of numbered references is created at the end of the main body—the system used herein. The complete bibliographic information for each reference is provided in the list of references, not in the main body. The references are cited in numerical order in the main body, not in a random pattern.

  As an alternative to using superscript numbers, the references may be cited in parentheses as "(Reference 1)," "(Reference 2)," etc., or abbreviated "(Ref. 1)," "(Ref. 2.)," etc., in the main body. If this is done, footnotes in the main body may be cited by superscript Arabic numbers rather than by superscript lowercase Roman letters (Chapter 2, Section 2.2.5).

- The author-date format places an author's surname and the year of the cited work in parentheses in the main body of the technical report; the rest of the bibliographic information is contained in a list of refences or bibliography. Many references in defense-related technical reports, e.g., military standards and specifications, do not have an author per se, so this system is not feasible.

- The number-identification system of citing works is similar in format to a footnote or endnote; however, the use of a footnote, endnote, or refenced link used in conjunction with a bibliography combines published and unpublished citations and does not make the distinction discussed previously in this section between a footnote or endnote and a cited formal reference.

Table 4-1 (beginning on page 181) lists example references that may appear in a defense-related technical report and indicates a suggested order of appearance of the items comprising a reference and a corresponding format based on the number-identification system. Table 4-1 lists the reference types alphabetically and is not necessarily comprehensive; references not listed should follow a similar order of appearance and format, as applicable.

ANSI/NISO Z39.18-2005 states the following regarding referenced figures and tables:

> If figures and tables are obtained from referenced material, the sources are identified in source or credit lines that are part of the figure(s) or table(s). A source or credit line contains adequate descriptive data to enable readers to verify the location of the original figure(s) or table(s). If the figure or table is used in its complete presentation (that is, both content and form), "Source" would be an appropriate lead-in to the citation. If either the content or form is modified, "Adapted from" would be appropriate lead-in wording. Such sources are not further identified in the list of references unless an additional reference to them appears in the [body] of the [technical] report.[2]

As discussed previously, figures and tables are comparable to the abstract and executive summary in their standalone nature. By including reference information in the figure or table, the reader is not required to refer to the list of references. Furthermore, figures and tables are oftentimes presented in different technical reports. The practice prescribed by ANSI/NISO Z39.18-2005 alleviates the need to repeat the citation from the list of references. If a figure or table contains reference information, the "source or credit lines"[2] should follow the arrangement and format indicated in Table 4-1. Also, if a figure or table is copyrighted, approval should be obtained from the originator.

Figures 4-9 and 4-10 show, respectively, a list of references based on the number-identification system in an unclassified and classified technical report using a typographical progression format. Figures 4-11 and 4-12 show, respectively, a list of references based on the number-identification system in an unclassified and classified technical report using a decimal numbering format.

As indicated in Table 4-1 and Figures 4-10 and 4-12, the only classified information that should appear in a list of references is the title of a document, if the title is classified. The rest of the citation should be unclassified, notwithstanding the overall classification of the document being cited. No classification mark should appear at the beginning of the citation.

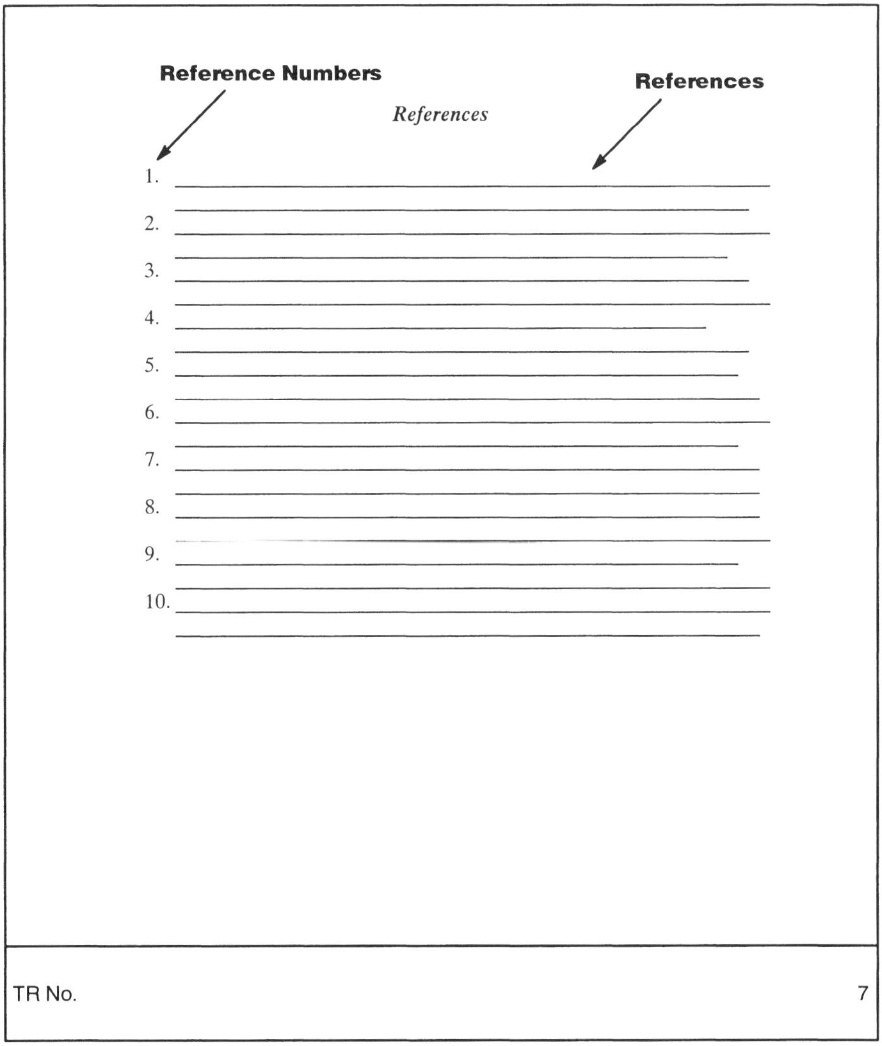

**Reference Numbers**                                        **References**

*References*

1. _____
   _____
2. _____
   _____
3. _____
   _____
4. _____
   _____
5. _____
   _____
6. _____
   _____
7. _____
   _____
8. _____
   _____
9. _____
   _____
10. _____
    _____

TR No.                                                              7

**Figure 4–9.**   Example List of References in
Unclassified Technical Report Using Typographical
Progression Format

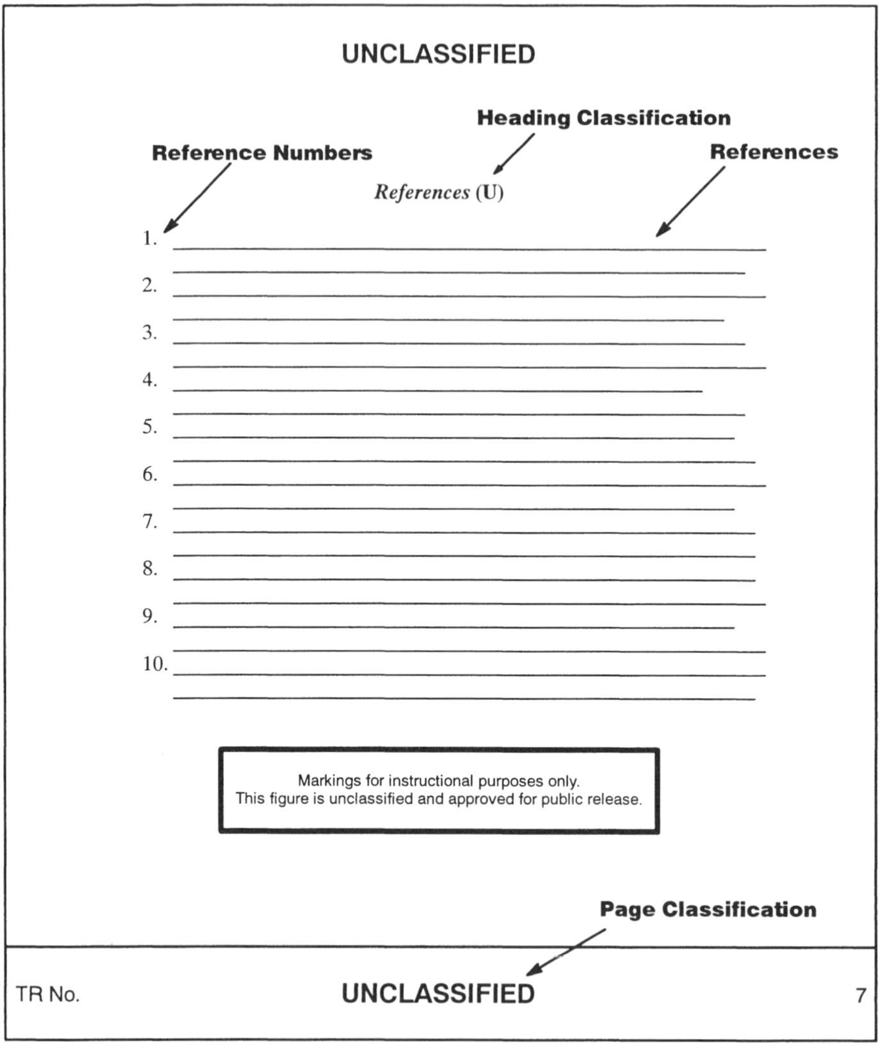

UNCLASSIFIED

**Heading Classification**

**Reference Numbers**

**References**

*References* (U)

1.

2.

3.

4.

5.

6.

7.

8.

9.

10.

Markings for instructional purposes only.
This figure is unclassified and approved for public release.

**Page Classification**

TR No.                    **UNCLASSIFIED**                    7

**Figure 4-10.** Example List of References in Classified
Technical Report Using Typographical Progression
Format

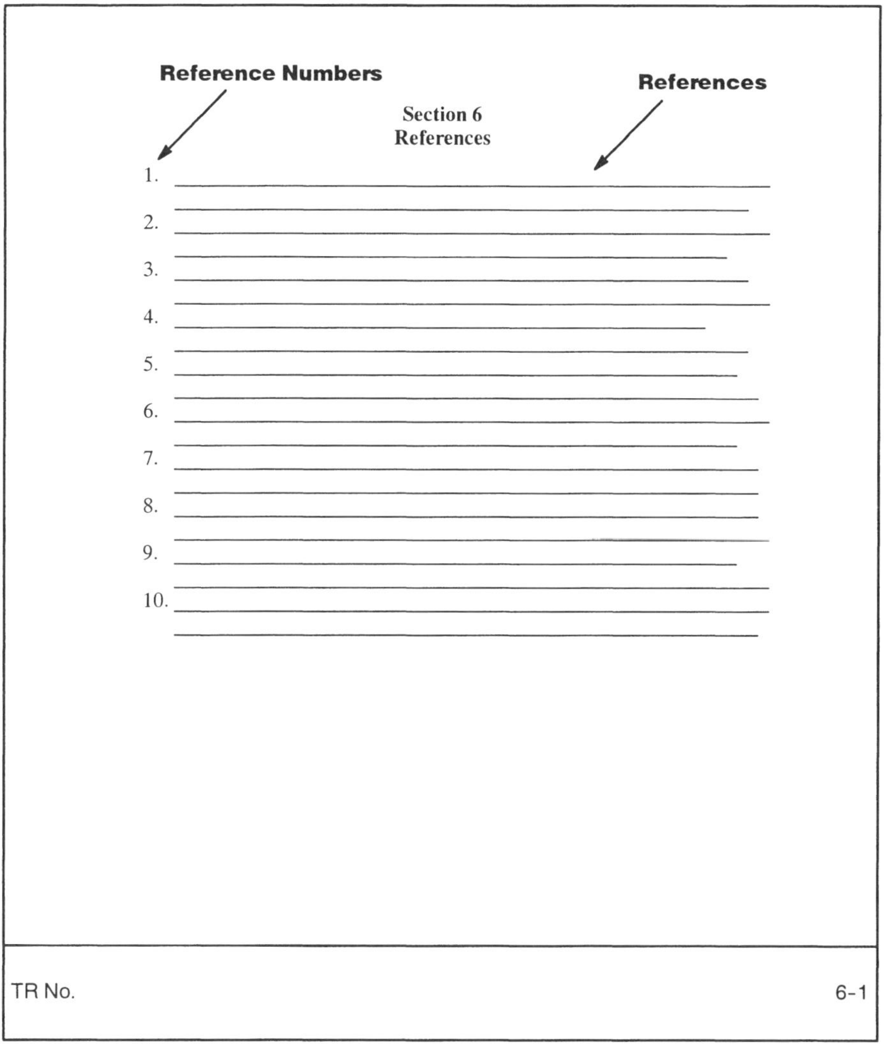

**Figure 4-11.**   Example List of References in
Unclassified Technical Report Using Decimal
Numbering Format

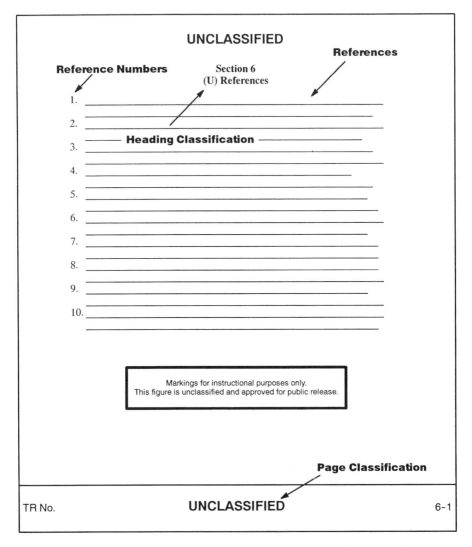

**Figure 4-12.**   Example List of References in Classified
Technical Report Using Decimal Numbering Format

## Table 4-1.　Example References

| Reference | Order of Appearance | Format |
|---|---|---|
| Book (Chapter/Pages) | 1. Author(s)[i]<br>2. Chapter Title<br>3. Book Title<br>4. Edition[ii]<br>5. Publisher Location<br>6. Publisher<br>7. Date<br>8. Page(s) Cited | Surname, First Initial. Middle Initial. (Principal Author), First Initial. Middle Initial. Surname (Co-Author), and First Initial. Middle Initial. Surname (Co-Author), "Chapter Title," in: *Book Title*, edition, Publisher Location: Publisher, Date, p. x or pp. x–x. |
| Book (Complete) | 1. Author(s)[i]<br>2. Title<br>3. Edition[ii]<br>4. Publisher Location<br>5. Publisher<br>6. Date | Surname, First Initial. Middle Initial. (Principal Author), First Initial. Middle Initial. Surname (Co-Author), and First Initial. Middle Initial. Surname (Co-Author), *Title*, edition, Publisher Location: Publisher, Date. |
| Book (Edited) | 1. Editor(s)[i]<br>2. Title<br>3. Edition[ii]<br>4. Publisher Location<br>5. Publisher<br>6. Date | Surname, First Initial. Middle Initial. (Principal Editor), First Initial. Middle Initial. Surname (Co-Editor), and First Initial. Middle Initial. Surname (Co-Editor), Eds., *Title*, edition, Publisher Location: Publisher, Date. |
| Computer Software (Classified) | 1. Software Name[iii]<br>2. Organization and Location<br>3. Operating System<br>4. Date[iv]<br>5. Classification[v] | "Software Name (U)," Organization and Location, Operating System, Date, CLASSIFICATION. |
| Computer Software (Unclassified) | 1. Software Name<br>2. Organization and Location<br>3. Operating System<br>4. Date[iv] | "Software Name," Organization and Location, Operating System, Date. |

[i]All co-authors/co-editors/co-inventors listed, if available. No space between initials.
[ii]If other than first edition. Abbreviate and lowercase, e.g., "2d ed."
[iii]Include classification.
[iv]Use military format, i.e., day month year, as applicable.
[v]All-capital letters. Include intelligence control marking (short form), if applicable.
[vi]Performing organization/sponsoring organization/DTIC/NTIS number(s), as applicable.
[vii]If applicable. Do not abbreviate.
[viii]Optional; required if public law not cited.
[ix]No space between initials.
[x]Abbreviate, e.g., Ph. D. or M.S. Dissertation.
[xi]Use acronym for organization.
[xii]Include protocol used, server identification, directory path, and file name.
[xiii]Includes other similar military/Government documents.
[xiv]If applicable.
[xv]Basic number only; omit revision-letter suffix.

## Table 4-1. Example References (Continued)

| Reference | Order of Appearance | Format |
|---|---|---|
| Conference Proceedings Article (Classified) | 1. Author(s)[i]<br>2. Article Title[iii]<br>3. Proceedings Title<br>4. Reference Number(s)[vi]<br>5. Proceedings Organization and Location<br>6. Date[iv]<br>7. Page(s) Cited<br>8. Classification[v] | Surname, First Initial. Middle Initial. (Principal Author), First Initial. Middle Initial. Surname (Co-Author), and First Initial. Middle Initial. Surname (Co-Author), "Article Title (U)," in: *Proceedings Title*, Organization Number(s), Proceedings Organization and Location, Date, p. x or pp. x-x, CLASSIFICATION. |
| Conference Proceedings Article (Unclassified) | 1. Author(s)[i]<br>2. Article Title<br>3. Proceedings Title<br>4. Reference Number(s)[vi]<br>5. Proceedings Organization and Location<br>6. Date[iv]<br>7. Page(s) Cited<br>8. Unclassified Controlled Information Marking[vii] | Surname, First Initial. Middle Initial. (Principal Author), First Initial. Middle Initial. Surname (Co-Author), and First Initial. Middle Initial. Surname (Co-Author), "Article Title," in: *Proceedings Title*, Organization Number(s), Proceedings Organization and Location, Date, p. x or pp. x-x, Unclassified Controlled Information Marking. |
| Congress, Act of | 1. Public Law Number<br>2. Title<br>3. U.S. Code (U.S.C.) Citation[viii]<br>4. Date[iv] | Public Law XX-XXX, "Title," XX U.S.C. §XXXX, Date. |
| Corporate-Authored Document (Classified) | 1. Corporate Author<br>2. Title[iii]<br>3. Reference Number(s)[vi]<br>4. Date[iv]<br>5. Classification[v] | Corporate Author, "Title (U)," Reference Number(s), Date, CLASSIFICATION. |

[i]All co-authors/co-editors/co-inventors listed, if available. No space between initials.
[ii]If other than first edition. Abbreviate and lowercase, e.g., "2d ed."
[iii]Include classification.
[iv]Use military format, i.e., day month year, as applicable.
[v]All-capital letters. Include intelligence control marking (short form), if applicable.
[vi]Performing organization/sponsoring organization/DTIC/NTIS number(s), as applicable.
[vii]If applicable. Do not abbreviate.
[viii]Optional; required if public law not cited.
[ix]No space between initials.
[x]Abbreviate, e.g., Ph. D. or M.S. Dissertation.
[xi]Use acronym for organization.
[xii]Include protocol used, server identification, directory path, and file name.
[xiii]Includes other similar military/Government documents.
[xiv]If applicable.
[xv]Basic number only; omit revision-letter suffix.

## Table 4–1.   Example References (Continued)

| Reference | Order of Appearance | Format |
|---|---|---|
| Corporate-Authored Document (Unclassified) | 1. Corporate Author<br>2. Title<br>3. Reference Number(s)[vi]<br>4. Date[iv]<br>5. Unclassified Controlled Information Marking[vii] | Corporate Author, "Title," Reference Number(s), Date, Unclassified Controlled Information Marking. |
| Defense Federal Acquisition Regulation Supplement (DFARS) | 1. Defense Federal Acquisition Regulation Supplement<br>2. Part/Subpart<br>3. Title | Defense Federal Acquisition Regulation Supplement, Part/Subpart Number, "Title." |
| Dissertation | 1. Author[ix]<br>2. Title<br>3. Dissertation Type[x]<br>4. University<br>5. Date[iv] | Surname, First Initial. Middle Initial., "Title," Dissertation Type, University, Date. |
| Drawing, Military (Classified) | 1. Organization[xi]<br>2. Drawing Number<br>3. Title[iii]<br>4. Date[iv]<br>5. Classification[v] | Organization Name, Drawing No. xxxx, "Title (U)," Date, CLASSIFICATION. |
| Drawing, Military (Unclassified) | 1. Organization[xi]<br>2. Drawing Number<br>3. Title<br>4. Date[iv]<br>5. Unclassified Controlled Information Marking[vii] | Organization Name, Drawing No. xxxx, "Title," Date, Unclassified Controlled Information Marking. |
| Executive Order | 1. Executive Order Number<br>2. Title<br>3. Date[iv] | Executive Order XXXXX, "Title," Date. |
| Federal Acquisition Regulation (FAR) | 1. Federal Acquisition Regulation<br>2. Part/Subpart<br>3. Title | Federal Acquisition Regulation, Part/Subpart Number, "Title." |

[i]All co-authors/co-editors/co-inventors listed, if available. No space between initials.
[ii]If other than first edition. Abbreviate and lowercase, e.g., "2d ed."
[iii]Include classification.
[iv]Use military format, i.e., day month year, as applicable.
[v]All-capital letters. Include intelligence control marking (short form), if applicable.
[vi]Performing organization/sponsoring organization/DTIC/NTIS number(s), as applicable.
[vii]If applicable. Do not abbreviate.
[viii]Optional; required if public law not cited.
[ix]No space between initials.
[x]Abbreviate, e.g., Ph. D. or M.S. Dissertation.
[xi]Use acronym for organization.
[xii]Include protocol used, server identification, directory path, and file name.
[xiii]Includes other similar military/Government documents.
[xiv]If applicable.
[xv]Basic number only; omit revision-letter suffix.

## Table 4–1.   Example References (Continued)

| Reference | Order of Appearance | Format |
|---|---|---|
| Internet Article | 1. Author(s)[i]<br>2. Title<br>3. Journal Name<br>4. Uniform Resource Locator (URL)[xii]<br>5. Accessed Date[iv] | Surname, First Initial. Middle Initial. (Principal Author), First Initial. Middle Initial. Surname (Co-Author), and First Initial. Middle Initial. Surname (Co-Author), "Title," *Journal Name*, http://www.x/x/x., Accessed: Date. |
| Journal Article (Classified) | 1. Author(s)[i]<br>2. Title[iii]<br>3. Reference Number(s)[vi]<br>4. Journal Name<br>5. Volume<br>6. Number<br>7. Date[iv]<br>8. Page(s) Cited<br>9. Classification[v] | Surname, First Initial. Middle Initial. (Principal Author), First Initial. Middle Initial. Surname (Co-Author), and First Initial. Middle Initial. Surname (Co-Author), "Title (U)," Organization Number(s), *Journal Name*, Vol. x, No. x, Date, p. x or pp. x-x, CLASSIFICATION. |
| Journal Article (Unclassified) | 1. Author(s)[i]<br>2. Title<br>3. Reference Number(s)[vi]<br>4. Journal Name<br>5. Volume<br>6. Number<br>7. Date[iv]<br>8. Page(s) Cited | Surname, First Initial. Middle Initial. (Principal Author), First Initial. Middle Initial. Surname (Co-Author), and First Initial. Middle Initial. Surname (Co-Author), "Title," Reference Number(s), *Journal Name*, Vol. x, No. x, Date, p. x or pp. x-x. |
| Military Instruction/ Directive (Classified)[xiii] | 1. Instruction/Directive Number<br>2. Subject[iii]<br>3. Date[iv]<br>4. Classification[v] | Instruction/Directive Number, "Title (U)," Date, CLASSIFICATION. |
| Military Instruction/ Directive (Unclassified)[xiii] | 1. Instruction/Directive Number<br>2. Subject<br>3. Date[iv]<br>4. Unclassified Controlled Information Marking[vii] | Instruction/Directive Number, "Title," Date, Unclassified Controlled Information Marking. |

[i]All co-authors/co-editors/co-inventors listed, if available. No space between initials.
[ii]If other than first edition. Abbreviate and lowercase, e.g., "2d ed."
[iii]Include classification.
[iv]Use military format, i.e., day month year, as applicable.
[v]All-capital letters. Include intelligence control marking (short form), if applicable.
[vi]Performing organization/sponsoring organization/DTIC/NTIS number(s), as applicable.
[vii]If applicable. Do not abbreviate.
[viii]Optional; required if public law not cited.
[ix]No space between initials.
[x]Abbreviate, e.g., Ph. D. or M.S. Dissertation.
[xi]Use acronym for organization.
[xii]Include protocol used, server identification, directory path, and file name.
[xiii]Includes other similar military/Government documents.
[xiv]If applicable.
[xv]Basic number only; omit revision-letter suffix.

## Table 4-1.    Example References (Continued)

| Reference | Order of Appearance | Format |
|---|---|---|
| Military Manual (Classified)[xiii] | 1. Manual Number<br>2. Title[iii]<br>3. Volume[xiv]<br>4. Part[xiv]<br>5. Revision[xiv]<br>6. Page(s) Cited<br>7. Date[iv]<br>8. Classification[v] | Manual Number, "Title (U)," Vol. X, Part X, Rev. X, p. x or pp. x–x, Date, CLASSIFICATION. |
| Military Manual (Unclassified)[xiii] | 1. Manual Number<br>2. Title<br>3. Volume[xiv]<br>4. Part[xiv]<br>5. Revision[xiv]<br>6. Page(s) Cited<br>7. Date[iv]<br>8. Unclassified Controlled Information Marking[vii] | Manual Number, "Title," Vol. X, Part X, Rev. X, p. x or pp. x–x, Date, Unclassified Controlled Information Marking. |
| Military Specification/ Standard (Classified)[xiii] | 1. Specification/Standard Number[xv]<br>2. Title[iii]<br>3. Date[iv]<br>4. Classification[v] | MIL-XXX-XXXXX, "Title (U)," Date, CLASSIFICATION. |
| Military Specification/ Standard (Unclassified)[xiii] | 1. Specification/Standard Number[xv]<br>2. Title<br>3. Date[iv]<br>4. Unclassified Controlled Information Marking[vii] | MIL-XXX-XXXXX, "Title," Date, Unclassified Controlled Information Marking. |
| Patent | 1. Inventor(s)[i]<br>2. Title<br>3. Patent Number<br>4. Reference Number(s)[vi]<br>5. Date[iv] | Surname, First Initial. Middle Initial. (Principal Inventor), First Initial. Middle Initial. Surname (Co-Inventor), and First Initial. Middle Initial. Surname (Co-Inventor), "Title," U.S. Patent No. x,xxx,xxx, Reference Number(s), Date. |

[i]All co-authors/co-editors/co-inventors listed, if available. No space between initials.
[ii]If other than first edition. Abbreviate and lowercase, e.g., "2d ed."
[iii]Include classification.
[iv]Use military format, i.e., day month year, as applicable.
[v]All-capital letters. Include intelligence control marking (short form), if applicable.
[vi]Performing organization/sponsoring organization/DTIC/NTIS number(s), as applicable.
[vii]If applicable. Do not abbreviate.
[viii]Optional; required if public law not cited.
[ix]No space between initials.
[x]Abbreviate, e.g., Ph. D. or M.S. Dissertation.
[xi]Use acronym for organization.
[xii]Include protocol used, server identification, directory path, and file name.
[xiii]Includes other similar military/Government documents.
[xiv]If applicable.
[xv]Basic number only; omit revision-letter suffix.

## Table 4-1.   Example References (Continued)

| Reference | Order of Appearance | Format |
|---|---|---|
| Standard (Industry) | 1. Publishing Organization<br>2. Title<br>3. Standard Number<br>4. Publisher Location<br>5. Publisher[xiv]<br>6. Date | Publishing Organization Name, "Title," Standard Number, Publisher Location: Publisher, Date. |
| Technical Report (Classified) | 1. Author(s)[i]<br>2. Title[iii]<br>3. Performing Organization<br>4. Technical Report Number(s)[vi]<br>5. Date[iv]<br>6. Classification[v] | Surname, First Initial. Middle Initial. (Principal Author), First Initial. Middle Initial. Surname (Co-Author), and First Initial. Middle Initial. Surname (Co-Author), "Title (U)," Performing Organization, Technical Report Number(s), Date, CLASSIFICATION. |
| Technical Report (Unclassified) | 1. Author(s)[i]<br>2. Title<br>3. Performing Organization<br>4. Technical Report Number(s)[vi]<br>5. Date[iv]<br>6. Unclassified Controlled Information Marking[vii] | Surname, First Initial. Middle Initial. (Principal Author), First Initial. Middle Initial. Surname (Co-Author), and First Initial. Middle Initial. Surname (Co-Author), "Title," Performing Organization, Technical Report Number(s), Date, Unclassified Controlled Information Marking. |
| Web Site | 1. Organization<br>2. URL[xii]<br>3. Accessed Date[iv] | Organization Name, http://www.x/x/x., Accessed: Date. |

[i]All co-authors/co-editors/co-inventors listed, if available. No space between initials.
[ii]If other than first edition. Abbreviate and lowercase, e.g., "2d ed."
[iii]Include classification.
[iv]Use military format, i.e., day month year, as applicable.
[v]All-capital letters. Include intelligence control marking (short form), if applicable.
[vi]Performing organization/sponsoring organization/DTIC/NTIS number(s), as applicable.
[vii]If applicable. Do not abbreviate.
[viii]Optional; required if public law not cited.
[ix]No space between initials.
[x]Abbreviate, e.g., Ph. D. or M.S. Dissertation.
[xi]Use acronym for organization.
[xii]Include protocol used, server identification, directory path, and file name.
[xiii]Includes other similar military/Government documents.
[xiv]If applicable.
[xv]Basic number only; omit revision-letter suffix.

# Chapter *5*

# *Back Matter*

ANSI/NISO Z39.18-2005 states that the back matter of a technical report "supplements and clarifies the body of the [technical] report (for example, appendices), makes the body easier to use (for example, glossary, lists of symbols, abbreviations, and acronyms, and index), and shows where additional information can found (for example, bibliography)."[2]

The back matter of defense-related technical reports contains only two required elements: the distribution list and the back cover. The remaining elements are either conditional or optional. If all of the elements are used, the back matter of defense-related technical reports should contain the following elements in the order presented:

a. Appendix(es)
b. Bibliography
c. List(s) of symbols, abbreviations, and acronyms
d. Glossary
e. Index
f. Distribution list
g. Back cover.

## 5.1 Appendixes

"Appendixes" (not "appendices") is the preferred plural form of "appendix" in Government spelling.[20] Appendixes are a conditional element and "contain information that supplements, clarifies, or supports the content [body]."[2] Appendixes are common in defense-related technical reports and are described as follows:

These conditional components [elements] of back matter also contain material that might otherwise interfere with an orderly presentation of ideas in the body. Placing detailed explanations, supporting data, or long mathematical analyses in appendices shortens the text [body] and makes it easier to read. However, information essential to the purpose of the [technical] report appears in the text [body].[2]

The last sentence is noteworthy because appendixes sometimes mistakenly contain information that should be included in the main body of the technical report.

Each appendix should be mentioned in the main body of the technical report, and the appendixes should appear in the order they are mentioned. Appendixes should be identified by letter, not by number; they should not be described as "attachments" or "addenda" or given other titles.

As stated previously, documents cited only in an appendix should not appear in the list of references, which is reserved for documents cited in the main body of the technical report; instead, they should appear as footnotes in the appendix, or if there are many references, a separate list of references may be created at the end of an appendix. Also, while the main body of the technical report refers to each appendix, an appendix should not refer back to the main body of the technical report, to the executive summary, or to another appendix. Basically, a one-way reference system exists, and each appendix is, in effect, a standalone document.

Appendixes may also comprise separate documents; however, previously published documents that would normally be cited and included in the list of references should not be appended, e.g., the full text of a technical report. The full text of certain documents, e.g., unpublished works, may be appended to a technical report if the appended material does not make the technical report unwieldy. When possible, only relevant portions of a separate document should be appended to a technical report.

Each appendix should begin on a new odd-numbered, right-hand page and should be identified by a capital letter, e.g., Appendix A, Appendix B, etc., and given a title. In a technical report using a typographical progression format, the numbering scheme should continue from the body of the technical report. In a technical report using a decimal numbering format, each appendix should be paginated according to its letter identification, e.g., A-1, A-2, etc.

In a technical report using a typographical progression format, if there is only one appendix, the appendix should not be lettered and should simply be referred to as "the appendix" in the main body of the technical report; only the title of the appendix should appear on the cover sheet or first page of the appendix. However, in a technical report using a decimal numbering format, each appendix should be identified by a capital letter, even if there is only one appendix.

Some separate appended documents are paginated beforehand and contain a table of contents and a list(s) of figures and tables. Unless impractical, the numbering scheme for the technical report should replace the original page numbers in a separate appended document. If the original numbering scheme is retained, the numbering scheme for the technical report should be added to the original page numbers with a slash separating the two, e.g., "5/A-5." Usually, only the appendix letter identification and title are included in the table of contents in the technical report, itself; appendix headings should not be included unless there is a compelling reason for their inclusion. Also, appendix figures and tables are usually not included in the list(s) of figures and tables in the technical report unless, again, there is a compelling reason for their inclusion.

In some instances, a cover sheet containing only the appendix letter identification and title is placed before the rest of the appendix. This is usually done when the appendix is a separate document; however, it also may be appropriate when the appendix comprises large figures or tables or other material requiring a separate cover sheet. The cover sheet is numbered. An unnumbered, left-hand blank page separates the cover sheet from the rest of the appendix but serves as the second page in the appendix. The appendix continues on a right-hand, odd-numbered page. This format applies to unclassified and classified appendixes.

Unclassified documents appended to classified technical reports do not need to have each portion marked unclassified; instead, each page of the appendix should be marked unclassified and the following statement should be included on the cover sheet or first page of the appended document indicating that the appendix is unclassified: "All portions of this appendix are unclassified."[5] Classified appendixes should be marked in the normal manner.

Figure 5-1 shows an example of an appendix in an unclassified technical report using a typographical progression format. Figure 5-2 shows an example of a cover sheet of an unclassified appendix in a classified technical report using a typographical progression format; Figure 5-3 shows an example of the body of the same appendix. Figure 5-4 shows an example of a classified appendix in a classified technical report using a typographical progression format.

Figure 5-5 shows an example of an appendix in an unclassified technical report using a decimal numbering format. Figure 5-6 shows an example of a cover sheet of an unclassified appendix in a classified technical report

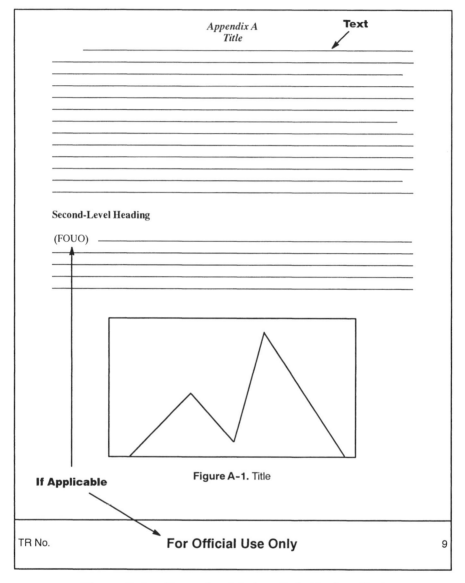

**Figure 5-1.** Example of Appendix in Unclassified Technical Report Using Typographical Progression Format

**Figure 5-1.**   Example of Appendix in Unclassified
Technical Report Using Typographical Progression
Format (Continued)

using a decimal numbering format; Figure 5-7 shows an example of the body of the same appendix. Figure 5-8 shows an example of a classified appendix in a classified technical report using a decimal numbering format.

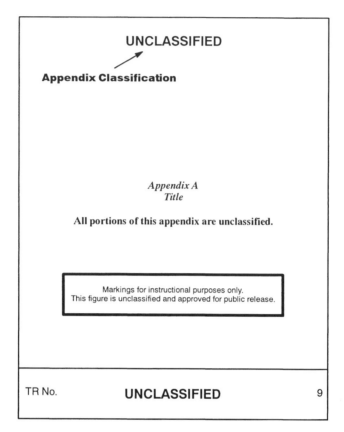

**Figure 5-2.**   Example of Cover Sheet of Unclassified
Appendix in Classified Technical Report Using
Typographical Progression Format

## 5.2    Bibliography

ANSI/NISO Z39.18–2005 states that a bibliography is a conditional element that is "include[d] when needed to amplify references."[2] The bibliography "lists additional sources of information not referenced in the text [body]."[2] However, because there is no specific condition upon which to include a bibliography—thus, making it exceptional—it is more appropriately considered an optional element in a defense-related technical report, especially if the technical report contains a list of references. A bibliography is a conditional element if there is no list of references and there is a need for a list of published works addressing the subject matter of the technical report and to aid in its comprehension, not merely a supplementary list of suggested readings.

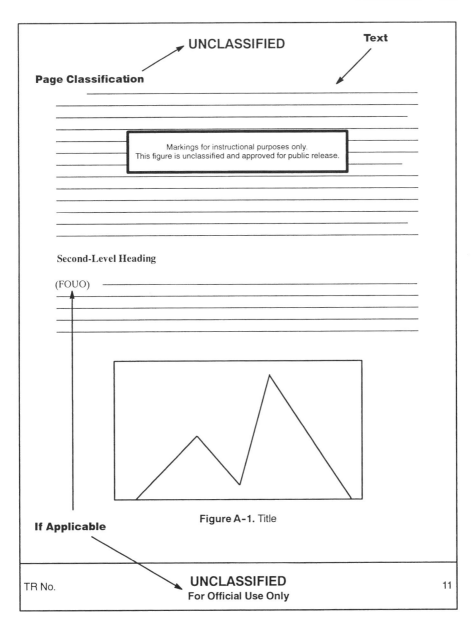

UNCLASSIFIED

Text

**Page Classification**

Markings for instructional purposes only.
This figure is unclassified and approved for public release.

Second-Level Heading

(FOUO)

Figure A-1. Title

**If Applicable**

TR No.

UNCLASSIFIED
For Official Use Only

11

**Figure 5-3.**   Example of Body of Unclassified
Appendix in Classified Technical Report Using
Typographical Progression Format

A bibliography, including its bibliographic entries, does not need to be referenced in the main body of the technical report. Unlike a list of references, bibliographic entries are not numbered as they would be in a num-

UNCLASSIFIED ◄——— **Page Classification**

**Second-Level Heading**

Markings for instructional purposes only.
This figure is unclassified and approved for public release.

**Table A-1.** Title

\*

†

\*
†

12                    **UNCLASSIFIED**                    TR No.

**Figure 5-3.**   Example of Body of Unclassified
Appendix in Classified Technical Report Using
Typographical Progression Format (Continued)

ber-identification system. As ANSI/NISO Z39.18–2005 states, "Bibliographic entries are usually arranged alphabetically by author/creator, but
any logical order may be used if it is explained and is consistent."[2] If the
bibliographic entries are arranged in alphabetical order and the entry con-

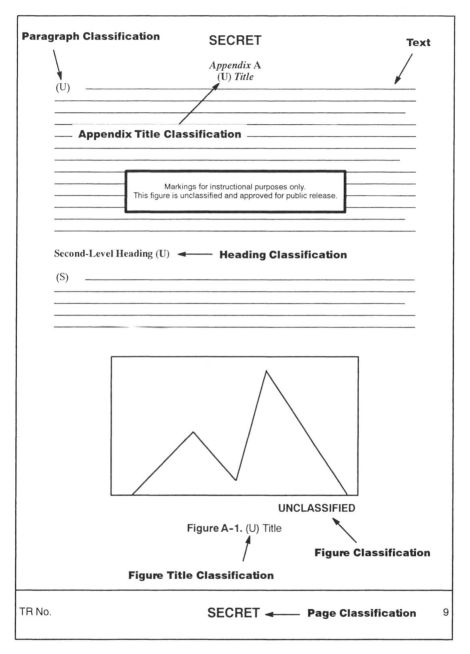

**Figure 5-4.** Example of Classified Appendix in Classified Technical Report Using Typographical Progression Format

CONFIDENTIAL ◄──── **Page Classification**

**Second-Level Heading (U)**

(U) _____

┌──────────────────────────────────────────┐
│ Markings for instructional purposes only. │
│ This figure is unclassified and approved for public release. │
└──────────────────────────────────────────┘

(C) _____

──────────────────── **Table Title Classification** ──

**Table A-1.** (U) Title

**Table Classification** ──► UNCLASSIFIED

10                    **CONFIDENTIAL**                    TR No.

**Figure 5-4.**   Example of Classified Appendix in
Classified Technical Report Using Typographical
Progression Format (Continued)

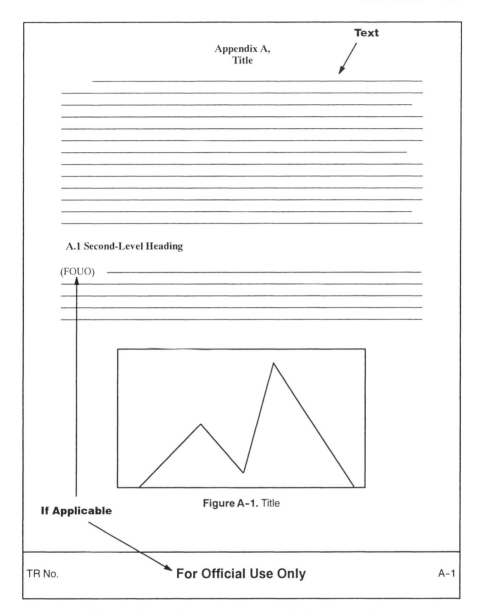

**Figure 5-5.**    Example of Appendix in Unclassified
Technical Report Using Decimal Numbering Format

tains a personal author, editor, or inventor, the surname and initials of the principal author, editor, or inventor are transposed: The surname should appear first, followed by a comma and the initials of the principal author, editor, or inventor. Co-authors, co-editors, and co-inventors should be ar-

**A.2 Second-Level Heading**

**Table A-1.** Title

A-2

TR No.

**Figure 5-5.**   Example of Appendix in Unclassified
Technical Report Using Decimal Numbering Format
(Continued)

ranged as they would be in a list of references. Other documents should also
be arranged as they would be in a list of references (Chapter 4, Table 4-1).
If a bibliographic entry begins with an indefinite or definite article, i.e., "a,"
"an," or "the," the next word should be considered in determining its order
of appearance.

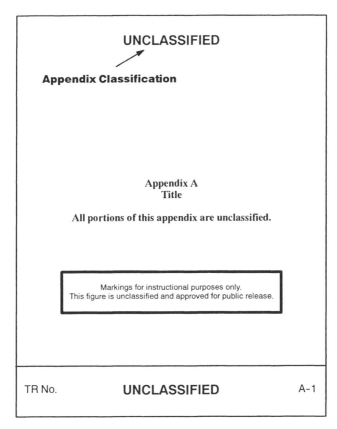

**Figure 5-6.**   Example of Cover Sheet of Unclassified Appendix in Classified Technical Report Using Decimal Numbering Format

In a classified technical report, the only classified information that should appear in a bibliography is the title of a bibliographic entry, if the title is classified. The rest of the bibliographic entry should be unclassified, notwithstanding the overall classification of the bibliographic entry. The page(s) and "Bibliography" should be marked unclassified.

A bibliography should begin on a new right-hand, odd-numbered page. In a technical report using a typographical progression format, the numbering scheme should continue from the body of the technical report or the appendixes. In a technical report using a decimal numbering format, the bibliography should be paginated as follows: "Bibliography-1," "Bibliography-2," etc., or abbreviated "Bib-1," Bib-2," etc. To avoid confusion with an appendix, a bibliography should not be paginated "B-1," "B-2," etc., in a technical report using a decimal numbering format. A suitable format for

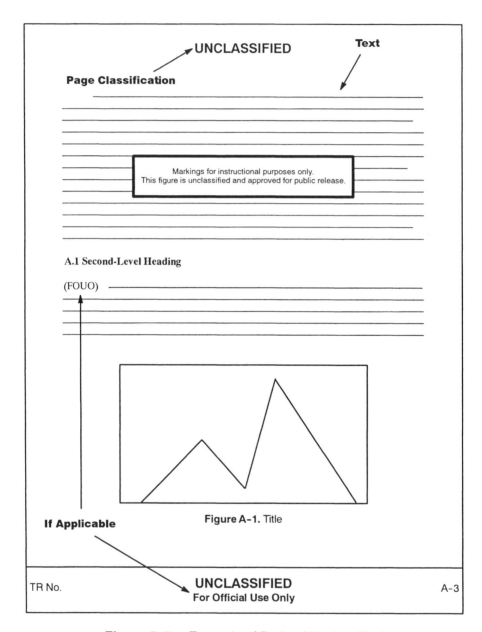

**Figure 5-7.**   Example of Body of Unclassified
Appendix in Classified Technical Report Using Decimal
Numbering Format

UNCLASSIFIED ◄──── **Page Classification**

A.2 Second-Level Heading

Markings for instructional purposes only.
This figure is unclassified and approved for public release.

**Table A-1.** Title

\*

†

\*
†

A-4                    **UNCLASSIFIED**                    TR No.

**Figure 5-7.**    Example of Body of Unclassified
Appendix in Classified Technical Report Using Decimal
Numbering Format (Continued)

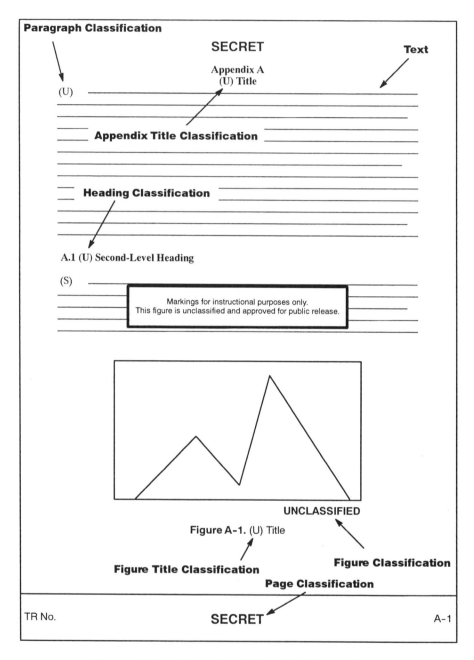

**Figure 5-8.**   Example of Classified Appendix in
Classified Technical Report Using Decimal Numbering
Format

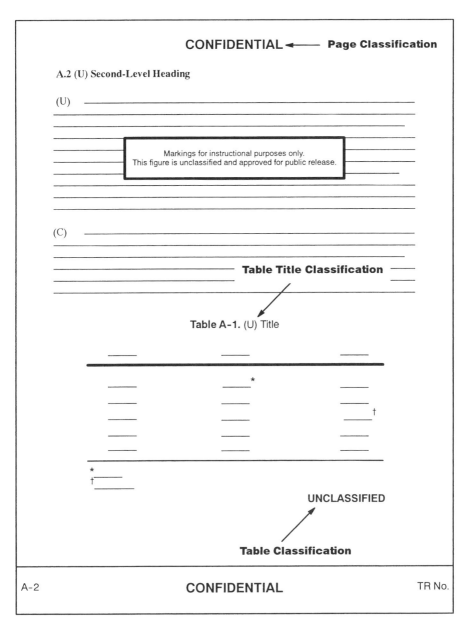

**Figure 5-8.** Example of Classified Appendix in
Classified Technical Report Using Decimal Numbering
Format (Continued)

each bibliographic entry is to begin each entry flush left and indent any additional lines 1 pica, referred to as a "hanging indentation."

Figures 5-9 and 5-10 show, respectively, a bibliography in an unclassified and classified technical report using a typographical progression format. Figures 5-11 and 5-12 show, respectively, a bibliography in an unclassified and classified technical report using a decimal numbering format.

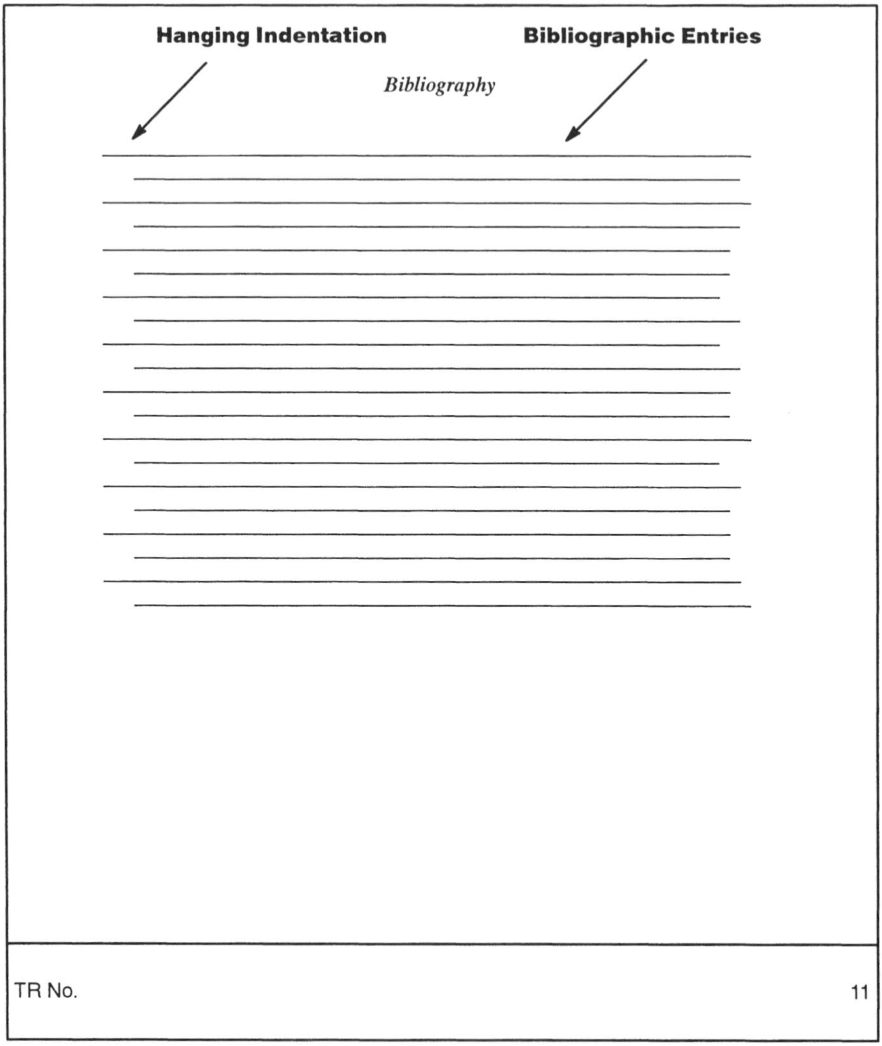

Figure 5-9. Example of Bibliography in Unclassified Technical Report Using Typographical Progression Format

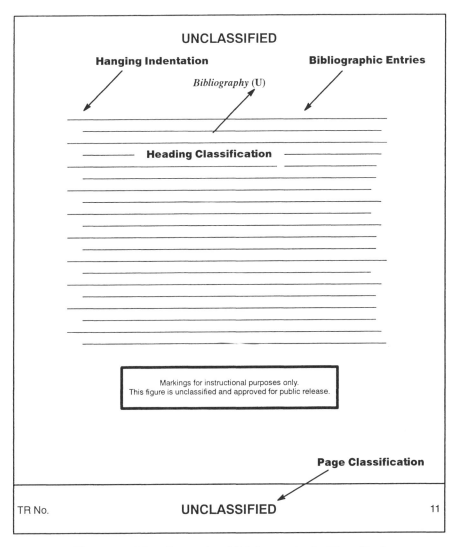

**Figure 5-10.** Example of Bibliography in Classified
Technical Report Using Typographical Progression
Format

## 5.3 List(s) of Symbols, Abbreviations, and Acronyms

ANSI/NISO Z39.18-2005 states that the list(s) of symbols, abbreviations, and acronyms is conditional and should be created if a technical report contains "more than five [symbols, abbreviations, and/or acronyms] that are not readily recognized as standard in the field."[2] It is highly probable that a defense-related technical report will contain enough acronyms alone to war-

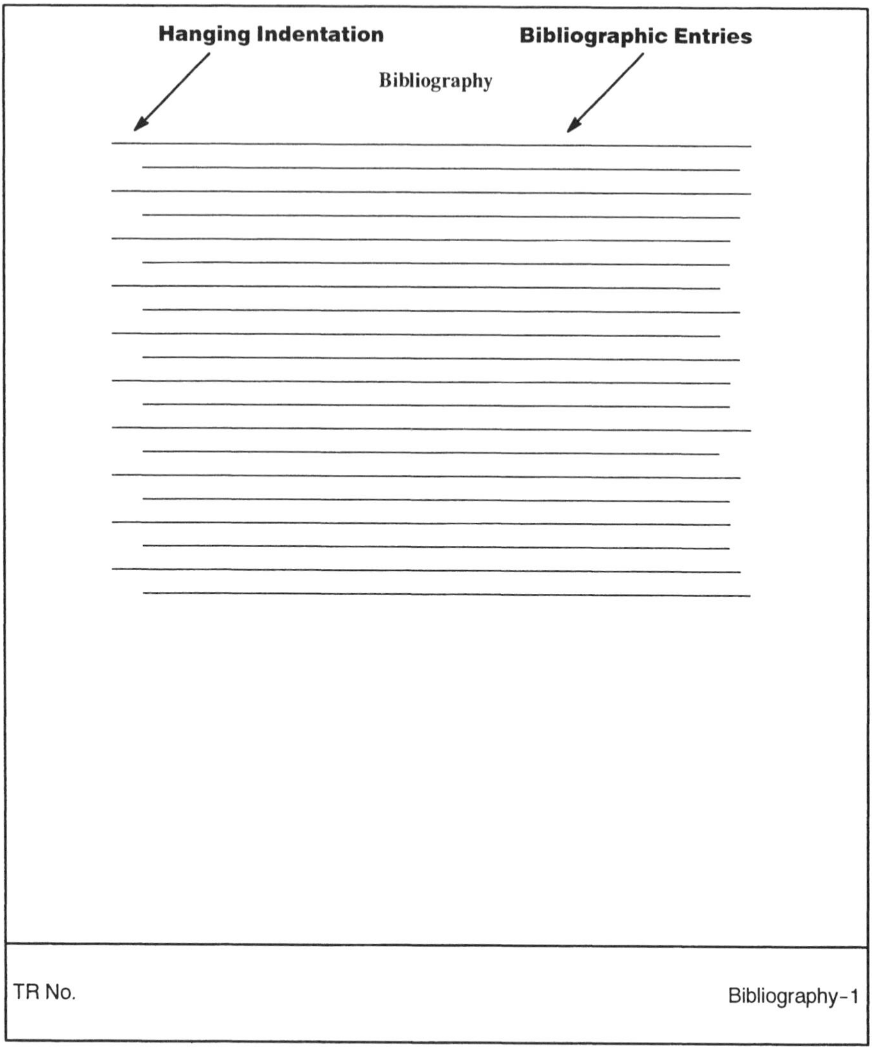

**Hanging Indentation**          **Bibliographic Entries**

Bibliography

TR No.                                                      Bibliography-1

**Figure 5-11.**   Example of Bibliography in Unclassified
Technical Report Using Decimal Numbering Format

rant such a list, making the list almost a requirement. Acronyms are considered abbreviations, and, if preferred, the title of this element may be shortened to "Symbols and Abbreviations" or, simply, "Abbreviations," if there are no symbols.

The purpose of the list is to provide the reader with a reference should the reader fail to retain the meaning of a symbol, abbreviation, or acronym. However, the list is not a substitute for defining all symbols, abbreviations,

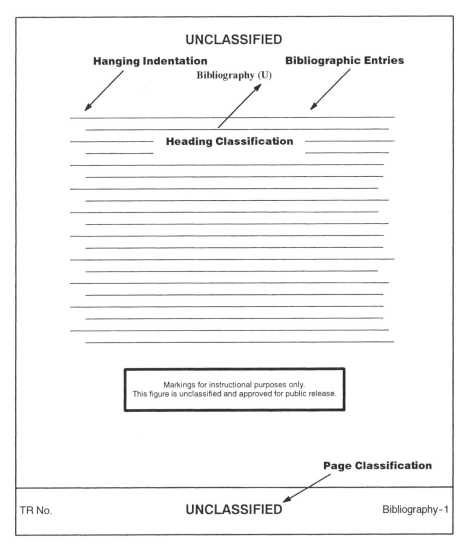

**Figure 5-12.**  Example of Bibliography in Classified
Technical Report Using Decimal Numbering Format

and acronyms at their first mention in all standalone elements and in the
main body of the technical report.

If each list is unusually long, the lists may be separate; however, in most
cases they may be combined into one alphanumeric list. If they are combined, the entries should appear in the following order, which is based, in
part, on ANSI/NISO Z39.18-2005[2]:

  a.  Arabic numerals
  b.  Roman numerals

   c.  Roman (English) alphabet capital letters
   d.  Roman (English) alphabet lowercase letters
   e.  Greek alphabet capital letters
   f.  Greek alphabet lowercase letters
   g.  Subscripts
   h.  Superscripts
   i.  Special notes.

ANSI/NISO Z39.18-2005 states, "If a symbol, abbreviation, or acronym has more than one definition, separate the explanations by a semicolon and explain each definition at its first use in the [technical] report."[2] Notwithstanding, to avoid confusion, a symbol, abbreviation, or acronym should have only one definition in a technical report. To alleviate this problem, a separate symbol, abbreviation, or acronym should be substituted for one of the two identical symbols, abbreviations, or acronyms. In the case of abbreviations and acronyms, no abbreviation or acronym should be used for one of the two identical abbreviations or acronyms; instead, the term should be spelled out in all instances. Conversely, two or more symbols, abbreviations, or acronyms should not be used to identify a single term unless it is unavoidable. For example, "DoD" and "DOD" are acronyms for the Department of Defense; both are recognized and accepted, and both are used interchangeably, especially in military standards and specifications. Acronyms within acronyms should be avoided.

Standard abbreviations for units of measurement, e.g., "lb" for pound, and common abbreviations, e.g., "U.S." for United States, do not need to be identified in standalone elements or in the main body of the technical report and should not be included in the list(s) of symbols, abbreviations, and acronyms.

The list(s) of symbols, abbreviations, and acronyms should include those symbols, abbreviations, and acronyms mentioned in the front matter, the body (exclusive of the list of references), and any appendixes in the technical report. However, if an appendix is a separate document and contains its own list(s) of symbols, abbreviations, and acronyms, these need not be repeated.

In classified technical reports, the list(s) of symbols, abbreviations, and acronyms should be unclassified. The page(s) and title(s) should be marked unclassified; however, no other classification markings should appear on the page(s).

The list(s) of symbols, abbreviations, and acronyms should begin on a new right-hand, odd-numbered page. In a technical report using a typographical progression format, the numbering scheme should continue from the body of the technical report, the appendixes, or the bibliography. In a technical report using a decimal numbering format, the list(s) of symbols, abbreviations, and acronyms should be paginated "SAA-1," "SAA-2," etc., or in a manner that would not confuse it with an appendix.

The list(s) of symbols, abbreviations, and acronyms should be arranged in two columns with the entry appearing on the flush left-hand side of the page and its corresponding definition aligned on the right-hand side of the page. If required, carryover lines in the corresponding definitions should be aligned flush left with the first line of the definition. The definition may appear in initial capital letters (used herein), whereby all words are capitalized except articles and prepositions of four or fewer letters, or appear as it would in the text. Ancillary sentences following the definition should be placed in parentheses.

Figures 5-13 and 5-14 show, respectively, a combined list of symbols, abbreviations, and acronyms in an unclassified and classified technical report using a typographical progression format. Figures 5-15 and 5-16 show, respectively, a combined list of symbols, abbreviations, and acronyms in an unclassified and classified technical report using a decimal numbering format. Appendix B, Section B.2.1.4 provides additional information regarding symbols, abbreviations, and acronyms.

## 5.4    Glossary

ANSI/NISO Z39.18-2005 states that a glossary is a conditional element "if a [technical] report incorporates terms unfamiliar to the intended audience."[2] However, like a bibliography, a glossary is exceptional and is more appropriately considered an optional element in defense-related technical reports. Much as a bibliography supplements a list of references, a glossary supplements a list(s) of symbols, abbreviations, and acronyms.

ANSI/NISO Z39.18-2005 states that "the glossary is a list of terms defined and explained to facilitate a reader's comprehension of the [technical] report when numerous terms requiring definition are used."[2] ANSI/NISO Z39.18-2005 adds that "glossary terms may also be defined at their first mention."[2]

As in a list(s) of symbols, abbreviations, and acronyms, a glossary incorporates those undefined terms mentioned in the front matter, the body (exclusive of the list of references), and any appendixes in the technical report. However, if an appendix is a separate document and contains its own glossary, these terms need not be repeated.

In classified technical reports, the glossary should be unclassified. The page(s) and title should be marked unclassified; however, no other classification markings should appear on the page(s).

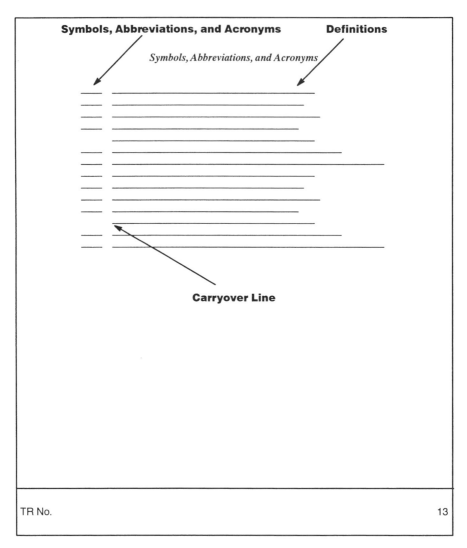

**Figure 5-13.** Example of List of Symbols, Abbreviations, and Acronyms in Unclassified Technical Report Using Typographical Progression Format

A glossary should begin on a new right-hand, odd-numbered page. In a technical report using a typographical progression format, the numbering scheme should continue from the body of the technical report, the appendixes, the bibliography, or the list(s) of symbols, abbreviations, and acronyms. In a technical report using a decimal numbering format, each appendix should be paginated as follows: "Glossary-1," "Glossary-2," etc., or abbreviated, "Gl-1," "Gl-2," etc., or in a manner that would not confuse it

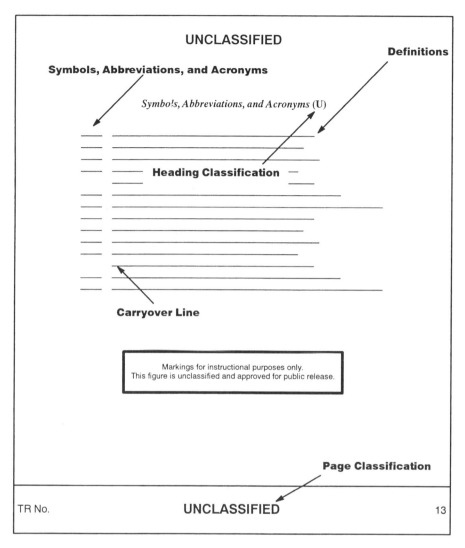

**Figure 5-14.**  Example of List of Symbols,
Abbreviations, and Acronyms in Classified Technical
Report Using Typographical Progression Format

with an appendix. As with a bibliography, a suitable format for each bibliographic entry is to use a hanging indentation of 1 pica. Each glossary term
should be in boldface type followed by a period. The first word in the definition should be capitalized, and the remaining words should be lowercase
unless they are proper nouns or adjectives. The definition may be a sen-
tence(s).

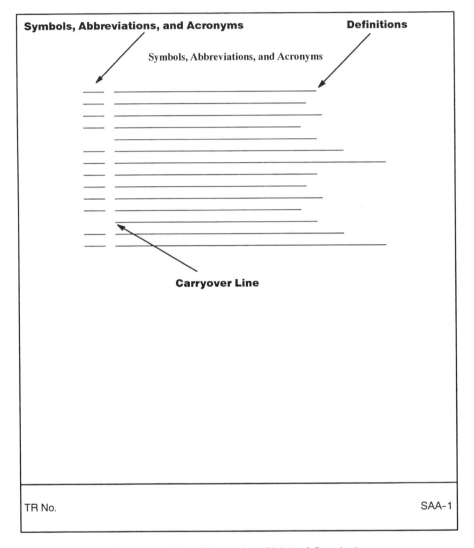

**Figure 5-15.**   Example of List of Symbols, Abbreviations, and Acronyms in Unclassified Technical Report Using Decimal Numbering Format

The glossary should appear in alphabetical order in two columns with the glossary term appearing on the flush left-hand side of the page and its corresponding definition aligned on the right-hand side of the page. Carryover lines in the corresponding definitions should be aligned flush left with the first line of the definition.

Figures 5-17 and 5-18 show, respectively, a glossary in an unclassified and classified technical report using a typographical progression format.

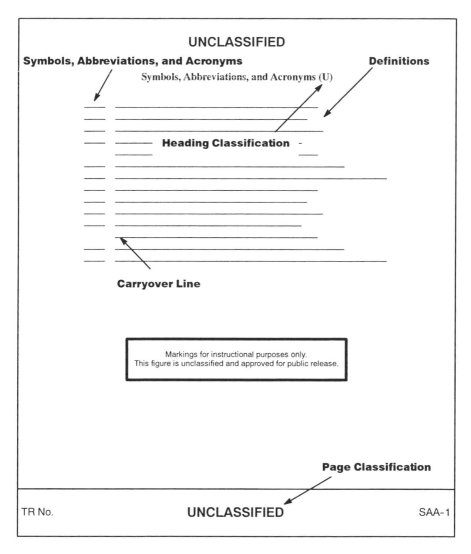

Markings for instructional purposes only.
This figure is unclassified and approved for public release.

**Figure 5-16.** Example of List of Symbols,
Abbreviations, and Acronyms in Classified Technical
Report Using Decimal Numbering Format

Figures 5-19 and 5-20 show, respectively, a glossary in an unclassified and classified technical report using a decimal numbering format.

## 5.5    Index

ANSI/NISO Z39.18-2005 states that "an index is an alphabetical listing of all major topics discussed in a [technical] report"[2] and is conditional and

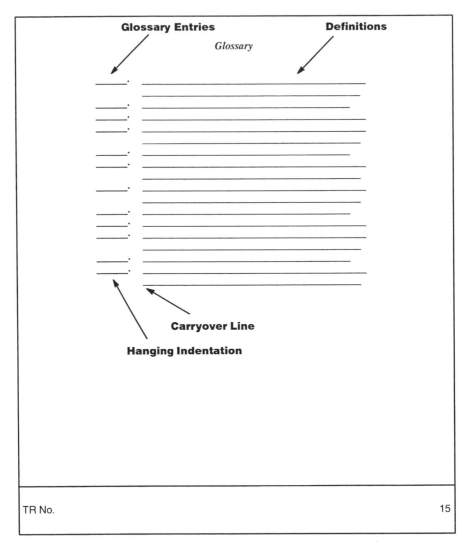

**Figure 5-17.** Example of Glossary in Unclassified
Technical Report Using Typographical Progression
Format

should be "include[d] when needed to ensure that a user locates all references to a concept."[2] However, an index, like a bibliography and a glossary,
is more appropriately considered an optional element in defense-related
technical reports. An index is rare in most technical reports because the
table of contents adequately performs the same function—albeit not in alphabetical order. However, in very large technical reports an index would be
helpful if not essential.

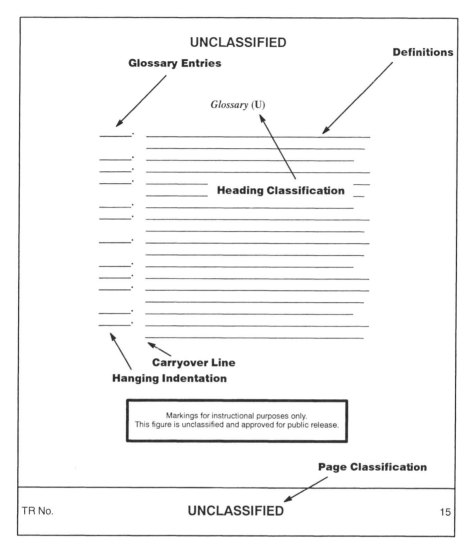

**Figure 5-18.** Example of Glossary in Classified
Technical Report Using Typographical Progression
Format

ANSI/NISO Z39.18-2005 mentions the various types of indexes: subject, name, number, and code[2]; however, in defense-related technical reports, a single, inclusive index should suffice. Names of individuals are usually de-emphasized in defense-related technical reports and, thus, should not be included in an index. This includes authors and those mentioned in the acknowledgments; however, organizational names may be in-

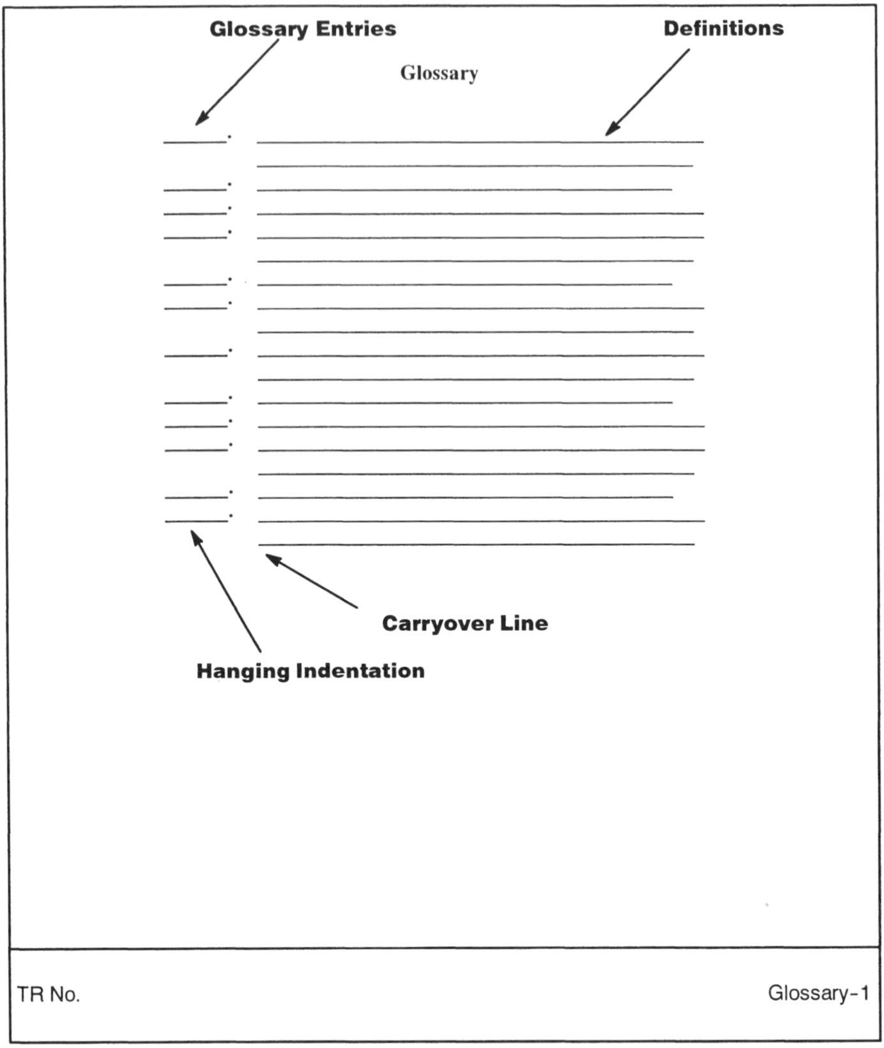

**Glossary Entries**                                    **Definitions**

Glossary

**Carryover Line**

**Hanging Indentation**

TR No.                                                    Glossary-1

**Figure 5-19.**   Example of Glossary in Unclassified
Technical Report Using Decimal Numbering Format

cluded. Most entries should be subject related. ANSI/NISO Z39.18–2005
adds the following:

> In preparing an index, the number and kind of access points (entry
> locations) and the information level of indexable matter (for example,
> abstract or concrete) are determined. Each index entry has a heading
> (first element) and a locator (page, section number, or linking informa-
> tion) where information about the entry is found. Terms used as [tech-

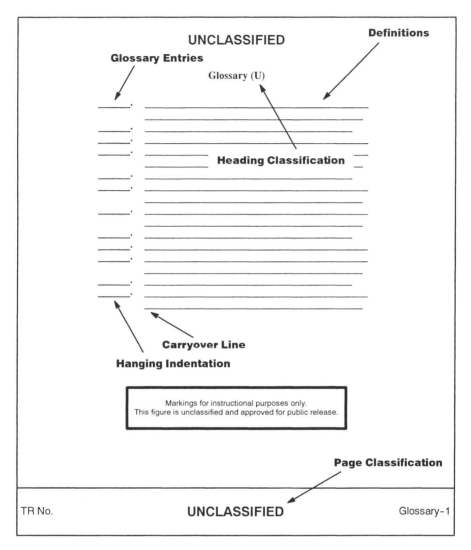

**Figure 5-20.**   Example of Glossary in Classified
Technical Report Using Decimal Numbering Format

nical] report headings are included in the index. The index contains all terms likely to be sought by the intended audience.[2]

An index should encompass those terms mentioned in the front matter (exclusive of the table of contents and the list[s] of figures and tables), the body of the technical report (exclusive of the list of references), and any appendixes in the technical report. However, if an appendix is a separate document and contains its own index, these entries need not be repeated.

To avoid having numerous page numbers after a single index entry, an entry should be divided into subcategories after five page numbers are assigned to the single entry; however, the entry should have a minimum of two subcategories before it is divided. Also, multiple-word entries should be logically transposed so that a reader can find an index entry at more than one location. (The index herein provides examples.) Again, after a transposed index entry has five page numbers assigned to it, a "see" reference should be placed at the transposed entries so that a reader is referred to one location for the index entry. "See also" references are also helpful for related entries.

Information indexed from figures and tables should be distinguished from information indexed from headings and text. The page numbers should be different in appearance, e.g., bold or italic, and a note at the beginning of the index or legend on each page should explain the difference.

In classified technical reports, the index should be unclassified. The page(s) and title should be marked unclassified; however, no other classification markings should appear on the page(s).

An index should begin on a new right-hand, odd-numbered page. In a technical report using a typographical progression format, the numbering scheme should continue from the body of the technical report; the appendixes; the bibliography; the list(s) of symbols, abbreviations, and acronyms; or the glossary. In a technical report using a decimal numbering format, the index should be paginated as follows: "Index-1," "Index-2," etc., or abbreviated, "Ind-1," "Ind-2," etc., or in a manner that would not confuse it with an appendix.

ANSI/NISO Z39.18-2005 states that each index entry should be lowercase "unless an entry begins with a proper name; indent entries uniformly for each level of modification. Indent runover lines deeper than the deepest entry."[2] As an alternative, main headings may be capitalized, and subheadings, if any, may be lowercase. Index entries should appear "single-spaced in a two- or three-column format."[2]

Figures 5-21 and 5-22 show, respectively, a two-column index in an unclassified and classified technical report using a typographical progression format. Figures 5-23 and 5-24 show, respectively, a two-column index in an unclassified and classified technical report using a decimal numbering format.

## 5.6    Distribution List

The distribution list is a required element in a defense-related technical report. It is the last element and should appear immediately before the back cover. Whereas a distribution statement indicates the limits, if any, of the

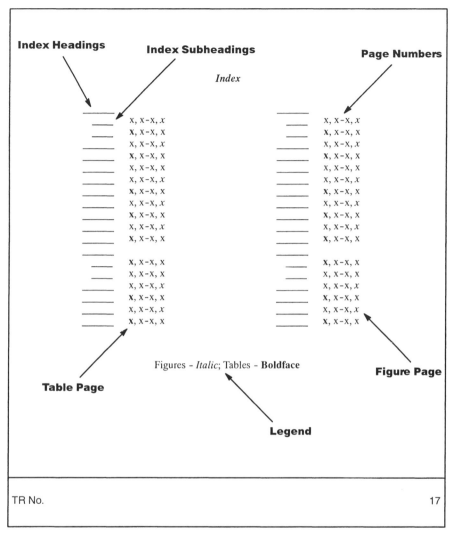

**Figure 5-21.** Example of Index in Unclassified
Technical Report Using Typographical Progression
Format

secondary distribution of a technical report, the distribution list indicates the initial or primary distribution of a technical report.

ANSI/NISO Z39.18-2005 states the following regarding the distribution list:

> The list indicates the complete mailing address of the individuals and organizations receiving copies of the [technical] report and the number

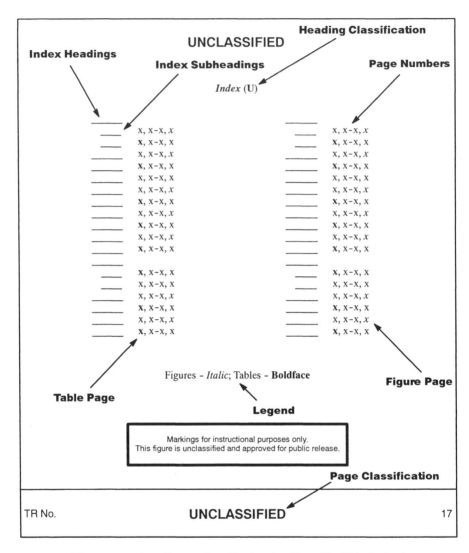

**Figure 5-22.** Example of Index in Classified Technical Report Using Typographical Progression Format

of copies received. The Privacy Act of 1974 forbids federal agencies from listing the names and home addresses of individuals, so in a government [technical] report a distribution list contains business addresses only.[2]

A distribution list is usually divided into an external list and an internal list. Those on the external list include individuals who work for other organizations, including sponsor representatives. Some are listed by organiza-

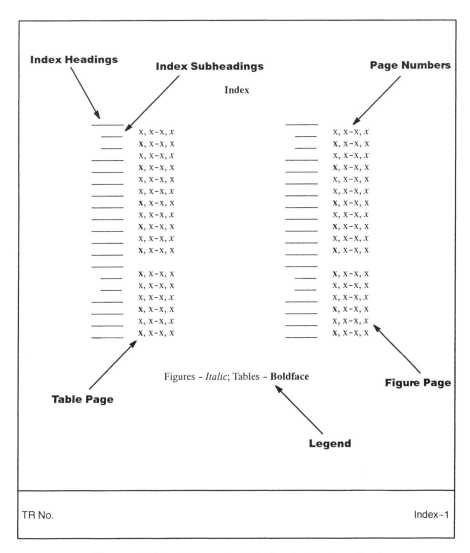

**Figure 5-23.**  Example of Index in Unclassified
Technical Report Using Decimal Numbering Format

tion and code alone without the use of surnames. The external list should also include DTIC. These individuals and DTIC routinely receive copies of technical reports from the performing organization, and they should be identified simply by their organization's acronym and their code and surname, as applicable. Occasionally, someone outside the normal routine receives a copy of a technical report, and they should be identified as indicated in ANSI/NISO Z39.18-2005.

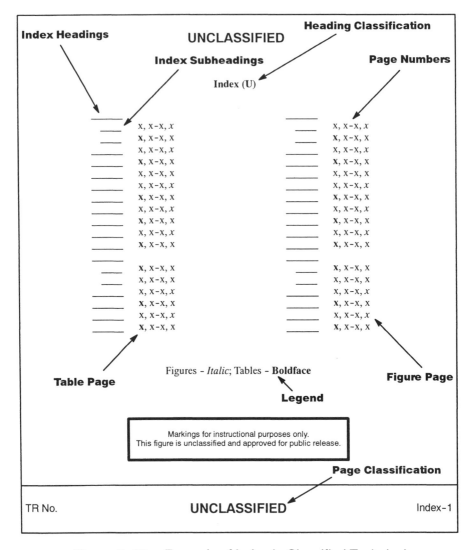

**Figure 5-24.** Example of Index in Classified Technical
Report Using Decimal Numbering Format

The internal list includes individuals who work in the performing organization and should be arranged in a hierarchical form, usually beginning with the department head and continuing through the author's immediate supervisor. The internal list also includes the author, co-authors, and other colleagues who might have an interest in the technical report and the organization's library or TIC.

The number of print copies each individual or organization receives is indicated in the distribution list. If more than one individual in an organization receives a copy, the organization should be listed first with the total number of copies it receives and each individual should be listed underneath the organization with the number of copies that individual receives. If an individual or organization receives an electronic copy of a technical report, this too should be indicated.

In classified technical reports, a serial number (Chapter 3, Section 3.6) should be assigned to each technical report, and this should be indicated in the distribution list in lieu of the number of copies an individual or organization receives. Electronic copies of classified technical reports should also be assigned a serial number.

ANSI/NISO Z39.18-2005 adds the following regarding distribution lists:

> In the case of classified [technical] reports, restricted-distribution [technical] reports, and [technical] reports containing proprietary data, such lists are extremely valuable as they can be used later for communicating instructions regarding handling and classification downgrading. A distribution list is also useful if errata [Chapter 2, Section 2.2.6] are discovered and changes are issued to correct a [technical] report.[2]

In classified technical reports, the distribution list should be unclassified. The page(s) and title should be marked unclassified; however, no classification markings should appear elsewhere on the page(s).

A distribution list should begin on a new right-hand, odd-numbered page. In a technical report using a typographical progression format, the numbering scheme should continue from the body of the technical report; the appendixes; the bibliography; the list(s) of symbols, abbreviations, and acronyms; the glossary; or the index. In a technical report using a decimal numbering format, the distribution list should be paginated as follows: "ID-1," "ID-2," etc., or in a manner that would not confuse it with an appendix.

Figures 5-25 and 5-26 show, respectively, a distribution list in an unclassified and classified technical report using a typographical progression format. Figures 5-27 and 5-28 show, respectively, a distribution list in an unclassified and classified technical report using a decimal numbering format.

## 5.7  Back Cover

The back cover is discussed with the front cover (Chapter 3, Section 3.3).

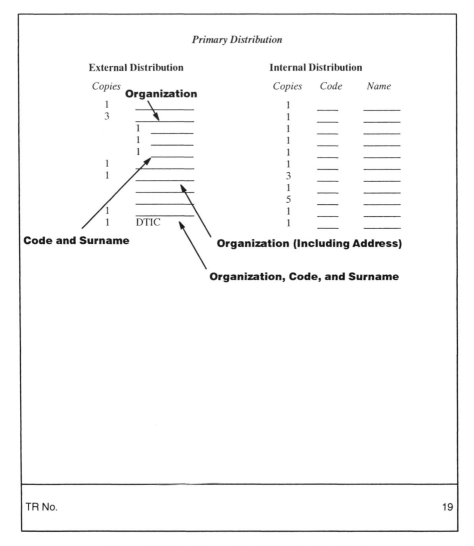

**Figure 5-25.**   Example of Distribution List in
Unclassified Technical Report Using Typographical
Progression Format

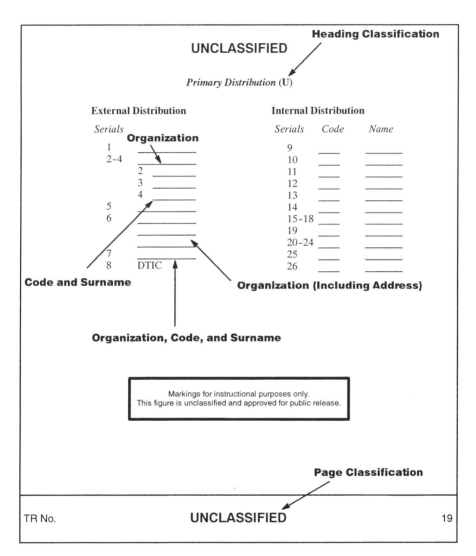

**Figure 5-26.**    Example of Distribution List in Classified
Technical Report Using Typographical Progression
Format

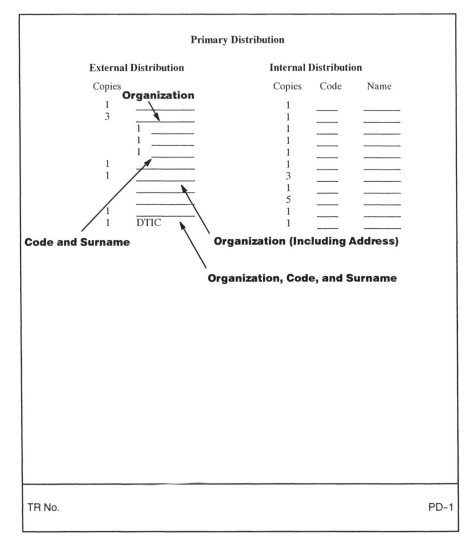

TR No.                                                                PD-1

**Figure 5-27.**   Example of Distribution List in
Unclassified Technical Report Using Decimal
Numbering Format

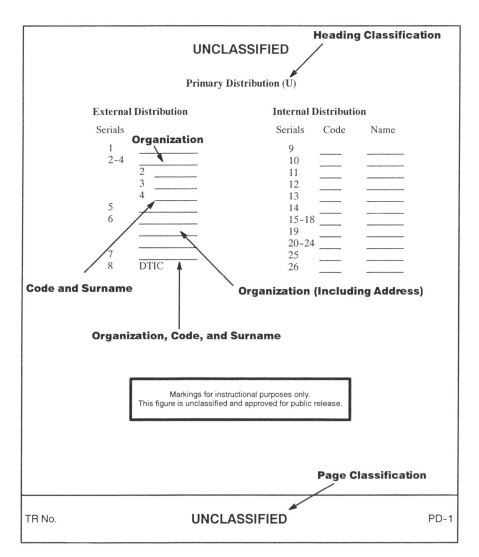

**Figure 5-28.** Example of Distribution List in Classified
Technical Report Using Decimal Numbering Format

# References

1. Department of Defense Instruction 4120.24, "Defense Standardization Program (DSP)," 18 June 1998.

2. National Information Standards Organization, "Scientific and Technical Reports - Preparation, Presentation, and Preservation," ANSI/NISO Z39.18-2005, Bethesda, MD: NISO, 2005.

3. Department of Defense Directive 5230.9, "Clearance of DoD Information for Public Release," 9 April 1996.

4. National Information Standards Organization, "Scientific and Technical Reports—Elements, Organization, and Design," ANSI/NISO Z39.18-1995, Bethesda, MD: NISO, 1995.

5. Cathcart, M.E., *STI Handbook: Guidelines for Producing, Using, and Managing Scientific and Technical Information in the Department of the Navy, A Handbook for Navy Scientists and Engineers on the Use of Scientific and Technical Information*, TD 2210, San Diego, CA: Naval Command, Control and Ocean Surveillance Center, RDT&E Division, 1 February 1992.

6. MIL-STD-38784, "Standard Practice for Manuals, Technical: General Style and Format Requirements," 2 July 1995.

7. Cathcart, M.E., *STI Handbook: Guidelines for Producing, Using, and Managing Scientific and Technical Information at NOSC*, Technical Document 1545, San Diego, CA: Naval Ocean Systems Center, September 1989.

8. MIL-STD-847, "Format Requirements for Scientific and Technical Reports Prepared by or for the Department of Defense," 7 November 1983.

9. American National Standards Institute, "Scientific and Technical Reports - Organization, Preparation, and Production," ANSI Z39.18-1987, New York, NY: ANSI, 1987.

10. National Information Standards Organization, "Understanding Metadata," http://www.niso.org/standards/resources/Understanding-Metadata.pdf, Bethesda, MD: NISO, 2004.

11. National Information Standards Organization, "Standard Technical Report Number Format and Creation," ANSI/NISO Z39.23-1997, Bethesda, MD: NISO, 1997.

12. National Information Standards Organization, "Guidelines for Abstracts," ANSI/NISO Z39.14-1997, Bethesda, MD: NISO, 1997.

13. DOD Instruction 3200.14, "Principles and Operational Parameters of the DoD Scientific and Technical Information Program," 13 May 1997.

14. DOD Directive 3200.12, "DoD Scientific and Technical Information (STI) Program (STIP)," 11 February 1998.

15. Department of Defense Financial Management Regulation, "Uniform Budget and Fiscal Accounting Classification," Volume 2B, Chapter 5, June 2004.

16. International Organization for Standardization, "Document management — Electronic document file format for long-term preservation — Part 1: Use of PDF 1.4 (PDF/A-1)," ISO 19005-1: 2005, Geneva, Switzerland: ISO, 2005.

17. Government-Industry Data Exchange Program (GIDEP), http://www.gidep.org, 7 July 2005.

18. Executive Order 13292, "Further Amendment to Executive Order 12958, As Amended, Classified National Security Information," 25 March 2003.

19. Secretary of the Navy Instruction 5510.36, "Department of the Navy (DON) Information Security Program (ISP) Regulation," 17 March 1999.

20. U.S. Government Printing Office, *United States Government Printing Office Style Manual 2000*, Washington: Superintendent of Documents, U.S. Government Printing Office, 2000.

21. Federal Acquisition Regulation, Subpart 52.227-14, "Rights in Data—General."

22. Defense Federal Acquisition Regulation Supplement, Subpart 227.7103-9, "Copyright."

23. Rice, W.W., "Environmental Life Cycle Management: An Overview of U.S. Navy and Other Department of Defense Initiatives," Naval Surface Warfare Center, Carderock Division, West Bethesda, MD 20084-5000, NSWCCD-63-TR—1999/15, September 1999.

24. Levedahl, W.J., S.R. Shank, and W.P. O'Reagan, "DD 21A—A Capable, Affordable, Modular 21st Century Destroyer," Naval Surface Warfare Center, Carderock Division, West Bethesda, MD 20084-5000, CARDIVNSWC-TR—93/013, December 1993.

25. DoD 5200.1-PH, "DoD Guide to Marking Classified Documents," April 1997.

26. DoD 5200.1-R, "Information Security Program," January 1997.

27. Department of Defense Directive 5400.7-R, "DOD Freedom of Information Act Program," September 1998.

28. Department of Defense, "A Reference Guide for Marking DoD Documents," AD-A 423 966.

29. Department of Defense Directive 5230.11, "Disclosure of Classified Military Information to Foreign Governments and International Organizations," 16 June 1992.

30. Department of Defense Directive 5210.2, "Access to and Dissemination of Restricted Data," 12 January 1978.

31. Office of the Under Secretary of Defense for Acquisition, Technology and Logistics, "Intellectual Property: Navigating Through Commercial Waters," Issues and Solutions When Negotiating Intellectual Property with Commercial Companies (Version 1.1), 15 October 2001.

32. Defense Federal Acquisition Regulation Supplement, Subpart 227.71, "Rights in Technical Data."

33. Defense Federal Acquisition Regulation Supplement, Subpart 227.72, "Rights in Computer Software and Computer Software Documentation."

34. Defense Federal Acquisition Regulation Supplement, Subpart 252.227-7013, "Rights in Technical Data—Noncommercial Items."

35. Defense Federal Acquisition Regulation Supplement, Subpart 252.227-7014, "Rights in Noncommercial Computer Software and Noncommercial Computer Software Documentation."

36. Defense Federal Acquisition Regulation Supplement, Subpart 252.227-7018, "Rights in Noncommercial Technical Data and Computer Software—Small Business Innovation Research (SBIR) Program."

37. Department of Defense Directive 5230.24, "Distribution Statements on Technical Documents," 18 March 1987.

38. Department of Defense Directive 5230.25, "Withholding of Unclassified Technical Data from Public Disclosure," 6 November 1984.

39. Government Accountability Office, "Defense Technologies: DOD's Critical Technologies Lists Rarely Inform Export Control and Other Policy Decisions," Highlights of GAO-06-793, A Report to the Committee on Armed Services, U.S. House of Representatives, July 2006.

40. Office of the Chief of Naval Operations Instruction 5510.161, "Withholding of Unclassified Technical Data from Public Disclosure," 29 July 1985.

41. DoD 5220.22-M, "National Industrial Security Program Operating Manual," Department of Defense, 28 February 2006.

42. Information Security Oversight Office, "Marking Classified National Security Information," May 2005.

43. Executive Order 12958, "Classified National Security Information," 17 April 1995.

Appendix *A*

# *Defense Technical Information Center (DTIC)*

## A.1    Overview°

DTIC is the premier provider of DoD STI, and it is the central point within DoD for acquiring, retrieving, and disseminating STI. DTIC contributes to the management and conduct of DoD RDT&E efforts by providing a method to access and transfer STI. DTIC serves as a vital link in the transfer of STI among DoD personnel, DoD contractors and potential contractors, and other Government agency personnel and their contractors. DTIC is a DoD field activity under the Under Secretary of Defense for Acquisition, Technology and Logistics, reporting to the Director, Defense Research and Engineering.

DTIC provides direct information support to the military. DTIC leverages the multibillion dollar investment in DoD scientific and technical research so that it can be used by the public, and it prevents unnecessary or redundant research from being performed at taxpayer expense.

The scope of DTIC's collection includes areas normally associated with defense research; however, because DoD's interests are widespread, the

---

°Unless indicated otherwise, the information in Appendix A was obtained from DTIC personnel or from DTIC's Web site: www.dtic.mil.

collection also contains information on various topics, including the following:

- Biology
- Chemistry
- Energy
- Environmental sciences
- Oceanography
- Computer sciences/information assurance
- Sociology
- Logistics
- Human factors engineering
- Manufacturing.

In addition to technical reports, DTIC also maintains the following holdings:

- Command histories
- Conference proceedings
- DoD directives and instructions
- DoD-sponsored patents and patent applications
- DoD-sponsored software and audiovisuals
- Foreign documents and translations
- Certain journal articles
- Management summaries
- Security classification guides
- Studies and analyses
- State-of-the-art tools
- Web resources.

DTIC has leading edge expertise in hosting Web sites and hosts approximately 100 Web sites for DoD components at the headquarters level. DTIC manages a number of Information Analysis Centers (IACs), which help customers locate, analyze, and use STI in a specialized subject area.

Anyone can search DTIC's publicly accessible collections and display or download STI, using the Public Scientific and Technical Information Network (STINET) service. DTIC also makes available unclassified, limited distribution information and classified information, including technical reports, to eligible users on the Private and Classified STINETs, respectively.

DTIC facilitates information exchange throughout the defense establishment to support the military. DTIC ensures technological innovations will continue to advance the military and the United States as a whole through the 21st century by maintaining an active collection program to acquire scientific, technical, and analytical information generated by the DoD community. DTIC encourages everyone in the defense community to participate in the program by submitting all applicable materials for inclusion in its technical reports database and research summaries collection.

DTIC contributors include DoD components and contractors working under DoD contract, as well as universities, non-profit organizations participating in scientific and technological activities for DoD, foreign governments, and Government agencies and their contractors. DTIC assigns a six-digit source code, which identifies and describes the contributor of documents to the technical reports database. This information includes the contributor's corporate name, subname, and address.

DTIC serves as DoD's system for secondary dissemination of technical reports. Documents submitted will always be available through DTIC; storage in DTIC is permanent. Submitting documents to DTIC reduces an organization's expenditures for information management, including costs of storage, printing, distribution of copies, maintaining documents on Web sites, and staffing of customer requests for documents. Technical reports at DTIC are easier to search and find because of the centralized database and the single search process used.

DFARS 235.010[p] and DFARS 252.235-7011[q] contain language requiring the submission of technical reports to DTIC. DoD Directive 3200.12[r] assigns DoD components responsibility for ensuring that DTIC is provided with pertinent material resulting from RDT&E programs. It also outlines their responsibilities when participating in DoD STI activities.

DTIC meets the FOIA requirement to have a public reading room and meets the requirements under the Anti-Deficiency Act to provide publicly releasable documents to the NTIS. NTIS is part of the Department of Commerce and serves as the largest central resource for Government-funded scientific-, technical-, engineering-, and business-related information. Copies of unclassified, unlimited distribution technical reports are forwarded by DTIC to NTIS. NTIS is responsible for selling Government publications to the general public. The Web site for NTIS is www.ntis.gov.

DTIC reduces the burden on an organization's legal or security staff in two ways: DTIC maintains a tracking and audit trail for limited and classified documents, and DTIC filters requests for release of documents to those

---

[p]Defense Federal Acquisition Regulation Supplement §235.010, "Scientific and Technical Reports."

[q]Defense Federal Acquisition Regulation Supplement §252.235-7011, "Final Scientific or Technical Report."

[r]DOD Directive 3200.12, "DoD Scientific and Technical Information (STI) Program (STIP)," 11 February 1998.

whose registration permissions are limited (DTIC Form 55 requests). Thus, an organization has to respond to far fewer FOIA requests.

Some authors are reluctant to provide DTIC with their technical reports because of their concern—however mistaken—that the information will not be properly safeguarded and may be distributed to individuals or organizations not authorized to receive it or they, simply, do not wish to inform DTIC of certain STI which they deem proprietary, i.e., they do not wish to share it. They even add notices to their technical reports stating that they should not be forwarded to DTIC. This practice is unwarranted and violates DoD policy, as noted in a DoD memorandum:

> A key product of the DoD Science and Technology program is knowledge. This product is made more valuable with reuse. Therefore, I strongly support and reaffirm DoD policy to aggressively pursue a coordinated and comprehensive Scientific and Technical Information Program (STIP) to meet the technological needs of the Warfighter, to enhance DoD's national security mission, and to provide the maximum contribution to the advancement of science and technology.

> As outlined in DoD Directive 3200.12, DoD Scientific and Technical Information Program, the Defense Technical Information Center (DTIC) is the central facility for the collection and dissemination of Science & Technology Information (STI) [sic] for DoD. DTIC makes this information available online to the Defense community, while controlling access according to security classification and distribution limitations. All information at DTIC is also preserved and protected through disaster recovery measures. A Science and Technology Portal to various STI databases is also being developed. The portal will allow the DoD R&D [research and development] community visibility into ongoing and emerging technologies.

> The DoD Research, Development, Test and Engineering (RDT&E) [sic] community is responsible for ensuring that DTIC is provided with material resulting from RDT&E programs including research that is only documented in journal articles, briefings and multimedia. Additionally, RDT&E investors are responsible for sending DTIC notification of changes to classification or distribution throughout the lifetime of the material. The URL for information on submitting information to DTIC is http:[//]www.dtic.mil.dtic/submitting.

> DTIC is able to electronically accept and process documents submitted in a wide variety of formats into the Technical Reports database. Accordingly, DoD RDT&E investigators should use electronic

submission to streamline the document submission process when possible.[s]

## A.2    DTIC Databases

DTIC maintains the following five databases:
*   Technical Report (TR)
*   Research Summaries (RS)
*   Independent Research and Development (IR&D)
*   Research and Development Descriptive Summaries (RDDS)

Figure A-1 shows the DTIC core collections (TR, RS, and IR&D databases) by level of access as of January 2005.[t]

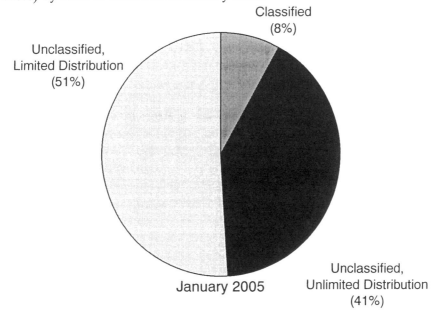

Figure A-1.   DTIC Core Collections by Level of
Access

---

[s]Sega, R.M., "Memorandum for Distribution, Subject: Submission and Dissemination of DoD Scientific and Technical Information," Director of Defense Research and Engineering, 3030 Defense Pentagon, Washington, DC 20301-3030, 4 April 2003.

[t]Sherman, H.Q., "NIPRNet . . . SIPRNet . . . How Do These Two Different Systems Affect Your Research?" www.dtic.mil/dtic/annualconf/conf05-NIPRNET-SIPRNET.ppt, 5 April 2005.

## A.2.1    Technical Report (TR) Database

### A.2.1.1    Overview

DTIC's TR database contains approximately 2 million citations in print and non-print format (software, datafiles, databases, and videorecordings) conveying the results of defense-sponsored RDT&E efforts. The TR database is inclusive and not limited strictly to technical reports; it also includes the following documents:

- Journal articles
- DoD-sponsored patent applications, studies, and analysis reports
- Open source literature from foreign countries
- Conference proceedings
- Reprints
- Theses.

DTIC assigns an accessioned document (AD) number to each item in the TR database. Each AD number is unique and is used to identify, control, and retrieve each item. DTIC's TR database contains only documents with the following security classifications: unclassified, foreign unclassified in confidence, foreign restricted, Confidential, and Secret. The DTIC TR database does not include Top Secret, cryptographic and communications security, and communications and electronic intelligence documents and does not accept these types of documents. Furthermore, DoD Directive 3200.12[u] excludes the DTIC collection of information relating to command and control of operations and operational forces, DoD scientific and technical intelligence production community products, and the DoD technical data management program.

### A.2.1.2    Submission Guidelines

DTIC prefers to receive electronic copies of technical reports as opposed to paper copies. In either case, only one copy is required. DTIC has the ability to accept and process all unclassified technical reports electronically into its TR database via the Internet using a registration and encryption process (Section A.4). Unclassified, unlimited distribution technical reports may be submitted via e-mail; however, unclassified, limited distribution and classified technical reports should not be submitted via e-mail.

Technical reports and portions thereof may be submitted in a wide variety of formats, including MS Word, Excel, PowerPoint, WordPerfect,

---

[u]DOD Directive 3200.12, "DoD Scientific and Technical Information (STI) Program (STIP)," 11 February 1998.

ASCII text, rich text, and PDF. If an unclassified technical report is not submitted electronically, one legible paper copy should be submitted via the U.S. Postal Service. Classified paper technical reports may also be submitted via the U.S. Postal Service, employing proper security safeguards. The mailing address is as follows:

ATTN: DTIC-O
Defense Technical Information Center
8725 John J. Kingman Rd.
Fort Belvoir, VA 22060-6218

In addition to distribution limitations, technical reports submitted to DTIC should be mailed as appropriate to their classification. If the technical report is classified (up to and including Secret), the technical report should be placed on appropriately marked removable media and mailed to DTIC, again using proper security safeguards.

DTIC ensures that the technical report is distributed only in accordance with the distribution statement and classification markings provided. DTIC will not accept a technical report without a proper distribution statement. DTIC also manages the systems for handling exceptions (DTIC Form 55). Complete instructions for submitting technical reports to DTIC are available on DTIC's Web site: www.dtic.mil/dtic/submitting/.

## A.2.2    Research Summaries (RS) Database

The RS database contains descriptions of DoD research that provide information on technical content, responsible individuals and organizations, principal investigators, and funding sources at the work unit level. Access to this collection is based on individual access restrictions. The collection consists of approximately 290,000 active and inactive summaries from 1965 to the present.

## A.2.3    Independent Research and Development (IR&D) Database

The IR&D database contains over 165,000 descriptions of R&D projects initiated by DoD contractors but not performed under contract. Federal statute encourages IR&D by allowing contractors to claim a portion of their IR&D costs as part of the overhead in "cost plus" contracts. By reimbursing these costs, DoD encourages industry to explore new technologies with potential application to military systems. IR&D projects are usually performed to improve existing DoD products, meet dual-use technology demands, or address a number of known or potential DoD requirements.

Accessible only to Government agencies, the information is used to identify contractors with expertise in areas of interest to DoD and to avoid DoD duplication of industry R&D efforts.

## A.2.4   Research and Development Descriptive Summaries (RDDS) Database

The RDDS database includes narrative information on RDT&E programs and project elements within DoD. The RDDS repository contains data from the military services and other DoD components. Its contents are derived from congressionally mandated reports.

## A.3   Information Analysis Centers (IACs)

IACs are research and analysis organizations chartered by DoD and managed by DTIC. Intended to help researchers, engineers, scientists, and program managers in Government, industry, and academia, IACs put expertise in the full spectrum of defense science and technology at the researcher's disposal. For fast answers (less than 4 hours) to questions and projects, IAC services are free. Extended projects (4 to 24 hours), searches and summaries (24 to 40 hours), reviews and analyses (40 to 80 hours), and technical area tasks (TATs) (more than 80 hours) are available at cost effective rates. TATs are separately funded work efforts over and above basic IAC products and services and are available only to the Government.

IACs improve the productivity of researchers, engineers, and program managers in the defense research, development, and acquisition communities by collecting, analyzing, synthesizing, and disseminating worldwide STI in clearly defined, specialized subject fields or subject areas. The IACs' secondary mission is to promote standardization within their respective fields. They accomplish these missions by providing in-depth analysis services and creating products. IACs respond to technical inquiries; prepare state-of-the-art reports, handbooks, and databooks; perform technology assessments; and support exchanges of information among scientists, engineers, and practitioners of various disciplines within the scope of a particular IAC.

IACs offer the following:
- Abstracts and indexes: announcements of pertinent reports in the IAC's field of interest;
- Bibliographic inquiry services: culled and authoritative bibliographic search reports;
- Critical reviews and technology assessments: the latest scientific and engineering information on specific technical subjects;

- Current awareness: newsletters and literature to promote subject area awareness;
- Referrals: consultation with or referral to worldwide recognized technical experts;
- Scientific and engineering reference works: useful and authoritative information in handbooks, databooks, and databases;
- State-of-the-art reports: summaries of the status of current technologies;
- TATs: detailed expert assistance in a wide range of technical support services; and
- Technical inquiry services: expert and authoritative advice in response to technical questions.

Currently, there are over 20 IACs; the exact number varies. DTIC manages most; the others are managed by other DoD components. The following subject areas are covered by the IACs:

- Advanced materials and processes
- Chemical warfare/chemical and biological defense
- Chemical propulsion
- Software
- Human systems
- Information assurance
- Infrared technology
- Modeling and simulation
- Manufacturing
- Nondestructive testing
- Reliability
- Survivability/vulnerability
- Weapon systems
- Military sensing technologies
- Airfields, pavements, and mobility
- Coastal engineering
- Cold regions science and technology
- Concrete technology
- Defense threat reduction
- Environment
- Hydraulic engineering
- Shock and vibration
- Soil mechanics.

The IAC Program Management Office (PMO) provides users with a central point of contact for the overall IAC program. The IAC PMO is the focal point for organizations requiring IAC services, contractors operating the respective IACs, and technical monitors overseeing projects and deliverables. The IAC PMO monitors IAC operations to ensure that their activities conform to the purposes and objectives of the DoD IAC program in general, and it provides continuous support to DoD and related Government technical communities. More information about the IACs may be found at the following Web site: iac.dtic.mil.

## A.4    Registration with DTIC

DTIC's collection includes information that describes research performed, sponsored, and co-sponsored by DoD on various topics. This information carries security classifications up the Secret level and is always marked by the originating office to show how it can be distributed. To ensure that the distribution of this sensitive information is not compromised, users are required to register with DTIC for access to most of DTIC's resources other than the Public STINET.

Based on the data submitted by potential users during the registration process and upon verification of those data, DTIC adds the user to its central registry file of authorized users. The registry file contains the user's level of security clearance. All requests for DTIC products and services are validated against the registry file for site classification and registration status.

DoD's STI is always marked by the organization that originates the document to show how it can be distributed. Many of the limits on distribution pertain to whether or not the user is a DoD employee, an employee of another Government agency, or a contractor for DoD or another Government agency. Because of these distribution limitations, access to DTIC data and DTIC's registration processes is different for different groups of users. The following are entitled to register with DTIC:

- Employees of DoD components (civilian and military, including the National Guard and reserves on active duty), DoD contractors, or potential DoD contractors;
- Employees of other Government agencies or Government contractors;
- Researchers from a university or college funded by DoD or a Government agency for conducting research throughout the United States;
- Participants in the Small Business Innovation Research or Small Business Technology Transfer program; and
- Faculty members, staff members, or students of Historically Black Colleges and Universities, Hispanic Serving Institutions, Tribal Colleges and Universities, or other minority institutions.

The registration process varies for the type of document and the type of applicant. If one does not fall into any of the previously listed categories, one can still search DTIC's unclassified, unlimited distribution collections on the Public STINET or order documents through NTIS.

## A.5    Research at DTIC

Research at DTIC may be performed online or by telephone. Documents may be located by using STINET, the Handle Service, or QuestionPoint.

### A.5.1    Scientific and Technical Information Network (STINET)

DTIC has three STINETs: the Public STINET, the Private STINET, and the Classified STINET.

#### A.5.1.1    Public STINET

The Public STINET contains only unclassified, unlimited distribution information. Anyone can access the Public STINET, and it is free of charge. No DTIC registration is required. The Public STINET provides access to citations to unclassified, unlimited distribution documents that have been entered into DTIC's technical reports database since December 1974, as well as some full text technical reports for those citations. The Public STINET also highlights featured material from the collection to make searches easier (special collections). The Public STINET includes links to many DoD information sources.

#### A.5.1.2    Private STINET

The Private STINET contains unclassified, unlimited distribution information and unclassified, limited distribution information and provides research summaries of ongoing research from 1965 to the present. Sanitized citations to classified technical reports are available. The Private STINET is free to eligible users; however, registration is required. The Private STINET is password protected. Access is based on individual validation. Documents may be downloaded free of charge, and electronic bibliographies are free. Paper documents may be ordered on the Private STINET for a fee. Research on the Private STINET is done on the Non-secure Internet Protocol Router Network (NIPRNet), a communications system for unclassified applications. The Private STINET includes the IR&D database of proprietary data, R&D descriptive summaries, and links to many other DoD information sources. The Private STINET is linked to the IACs through the Total Electronic Migration System, which contains the collections of the IACs.

### A.5.1.3   Classified STINET

The Classified STINET contains unclassified, unlimited distribution information; unclassified, limited distribution information; and classified information up to the Secret level. Registration is required for the Classified STINET. Research on the Classified STINET is done on the Secret Internet Protocol Router Network (SIPRNet), a communications system for classified applications. Before registering for the Classified STINET, a DTIC user should have the following: a Secret clearance (minimum), a SIPRNet workstation, and a SIPRNet e-mail address. The Classified STINET contains the IR&D database of proprietary data, unclassified and classified R&D descriptive summaries, and links to many other DoD information sources.

### A.5.2   Handle Service

DTIC's Handle Service assigns pertinent names called "handles" to electronic resources and stores them in its Handle Service. The correct location (URL) where the resource can be found is stored with each handle. When the location changes, the handle record is updated to reflect this change, and there is no broken link. DTIC's Handle Service facilitates the preservation and long-term access to electronic resources, ensuring their availability over time. DTIC's Handle Service plays a vital role in preserving DoD's Internet-accessible resources.

### A.5.3   QuestionPoint ("Ask a Librarian")

QuestionPoint or "Ask a Librarian" is a responsive, digital, collaborative reference service available 24 hours a day, 7 days a week. It allows libraries or TICs to expand reference services with shared resources and subject specialists worldwide. DTIC is a participating member of the following QuestionPoint cooperatives: the Global Reference Network, a worldwide group of libraries and institutions committed to digital reference, and the Defense Digital Library Research Service (DDLRS), a 24-hour-a-day, 7-day-a-week electronic reference for DoD libraries. Questions submitted to QuestionPoint may be referred to the Global Reference Network or the consortium of DoD libraries for a response.

*Appendix* **B**

---

# *Tone and Style*

## B.1  Tone

Alliteratively speaking, technical reports should be clear, concise, consistent, and correct. While a personal, informal tone may be acceptable in other forms of technical communication, an impersonal, formal tone, characterized by its objectivity, should be used in preparing defense-related technical reports. To help achieve this, certain writing and related practices should be followed. The following list addresses some of these practices.

- Good technical reports should, like literary short stories, be able to be read at one sitting. Less is usually better than more in technical report writing. Time is precious. The longer a technical report is, the less likely it is to be read in its entirety, if at all. If it is necessary to have a lengthy technical report (more than 50 pages), particular emphasis should be placed on creating a thorough executive summary and appending or eliminating non-essential information.
- Defense-related technical reports should be written to a college-level audience with a general scientific and technical background. Esoteric writing should be avoided or kept to a minimum. This is especially true when writing the abstract and executive summary.
- While scientific papers may be written in the first person, defense-related technical reports should be written in the third person. First-person pronouns such as "I" and "we" should not appear in a technical report. The personal author represents or performs work on behalf of the corporate author (performing organization), which is ultimate-

ly responsible for the work. As such, the personal author "speaks" for the corporate author, and the technical report should be written to indicate this relationship. While a defense-related technical report should be written in the third person, the name(s) of the personal author should not be mentioned in the body of the technical report and third-person references to "the author" should not appear. Second-person references to the reader, direct or implied, are also inappropriate.

- Generally, the active voice is preferred over the passive voice in writing. However, in defense-related technical report writing, it is sometimes necessary to write in the passive voice because of the need to write in the third person. Inserting the name of the corporate author as the subject of a sentence in the active voice can oftentimes be awkward, e.g., "The [corporate author] reached several conclusions." This can be corrected by using the passive voice, thereby emphasizing the object or receiver of the action and omitting the implied actor or doer (corporate author), e.g., "Several conclusions were reached."

- Emphatic phrases such as "it should be noted" should be avoided. They, generally, add nothing to a sentence and tend to diminish other information that is not emphasized as such. All information in a technical report is noteworthy; otherwise, it should not be included.

- The use of italic and/or boldface type or underlining for emphasis in a technical report should be avoided. Emphasizing certain aspects of a technical report, in effect, de-emphasizes the rest of the text. (Italic and boldface may be used to distinguish computer commands from the rest of the text.)

- The terminology of one's discipline may be used; however, the jargon of one's service or laboratory should be avoided, e.g., "toilet and washroom," not "head." In defense-related technical reports, no particular scientific or engineering discipline should be unduly emphasized, regardless of subject matter. Modifications to existing words resulting in "new" undefined terms are inappropriate; only defined terms should be used.

- Sexually biased or gender-oriented language should be avoided. Pronouns such as "he" and "she" can be replaced by pluralizing the antecedent (the word the pronoun refers to), pluralizing pronouns while leaving antecedents singular, or otherwise rephrasing the sentence. Sex references in job titles, e.g., crewman; male and female gender word forms, e.g., aviatrix; and stereotyped terms or expressions, e.g., man-hour, should also be avoided.

- While abbreviations and symbols are acceptable, word contractions should not appear in technical reports. Therefore, "it's," a contraction for "it is," should never appear in a technical report. "Its" is the possessive form of "it."

- All-capital letters are reserved almost exclusively for acronyms and should not be used for ships, missiles, and other weapons systems, e.g., "Aegis cruiser," not "AEGIS cruiser."

- Annotated viewgraph compilations are byproducts of audiovisual presentations. In annotated viewgraph compilations, the text complements the figures and tables; in technical reports, the figures and tables complement the text. Viewgraph compilations should not be presented as technical reports and vice versa.

- Questions should be avoided in technical reports, especially rhetorical questions. Technical reports should provide answers, not pose questions.

- Some results of RDT&E may be exciting; however, exclamatory sentences, i.e., those containing an exclamation point, should be avoided.

- While redundancy is an inherently good feature of many military systems, it should be avoided in technical reports. A topic only needs to be discussed once in the main body of the technical report and should only appear in its proper element. For example, conclusions and recommendations should only appear in their respective elements within the technical report, not inserted randomly throughout the technical report.

- Ships (and other military hardware) are inanimate objects and should be described as such. The definite article "the" should precede the name of a ship. For example, "The *Ronald Reagan* returned to port," not "*Ronald Reagan* returned to port." In naval tradition, ships are often personified by assigning feminine gender characteristics to them, e.g., "sister ship." This may be acceptable in intraservice communications; however, defense-related technical reports are read by civilians and others outside the military who may not be familiar or comfortable with naval expressions.

- When writing about the U.S. Armed Forces, the name of the service should be identified in its entirety at its first mention in the abstract, executive summary, and/or the main body of the technical report, e.g., "the U.S. Army." Thereafter, it may be referred to, simply, as "the Army" unless a foreign army is mentioned in the technical report; if so, it should always be referred to as "the U.S. Army."

- Shortened, informal, or slang terms, e.g., "spec" for "specification," "intel" for "intelligence," and "helo" for "helicopter," should be avoided. A term should always be spelled out. "Get" should also be avoided.

## B.2   Style

Authors should adhere to accepted publishing practices in preparing defense-related technical reports. These practices include incorporating a formal style into their writing. Formal style is usually documented to promote consistency in presentation.

### B.2.1   United States Government Printing Office Style Manual

The *United States Government Printing Office Style Manual*[v] is the official style manual of the U.S. Government and, therefore, DoD. Questions regarding word capitalization, spelling, compounding, punctuation, abbreviations and letter symbols, signs and symbols, italic, and numerals are addressed in the *United States Government Printing Office Style Manual*. ANSI/NISO Z39.18–1995[w] is referenced in the *United States Government Printing Office Style Manual*.

The *United States Government Printing Office Style Manual* should, in general, be followed in matters relating to style in preparing defense-related technical reports. Occasionally, deviation from the *United States Government Printing Office Style Manual* is required; however, this should be noted in an individual organizational style manual and kept to a minimum to avoid inconsistencies within and among defense-related technical reports.

Applicable portions of the *United States Government Printing Office Style Manual* are repeated in Sections B.2.1.1 to B.2.1.5 for convenience; some portions have been modified and consolidated for clarification and to include additional examples and exceptions.

The *United States Government Printing Office Style Manual* can be downloaded and printed free of charge at the following Web site: www.gpo.gov. or purchased through the Government Printing Office.

---

[v]U.S. Government Printing Office, *United States Government Printing Office Style Manual 2000*, Washington: Superintendent of Documents, U.S. Government Printing Office, 2000.

[w]National Information Standards Organization, "Scientific and Technical Reports—Elements, Organization, and Design," ANSI/NISO Z39.18–1995, Bethesda, MD: NISO, 1995.

## B.2.1.1    Editor's and Illustrator's Marks

Authors are oftentimes required to interact not only with other engineers and scientists but with editors and illustrators and vice versa. Publishing notation, a form of shorthand, enhances communication among the disciplines, thereby expediting completion of a technical report. Figure B–1 shows the standard editor's and illustrator's marks used to indicate changes to a technical report.

## B.2.1.2    Capitalization

Capitalization is frequently an issue in defense-related technical reports. Excessive capitalization should be avoided. As a rule, common nouns and adjectives should appear lowercase, and proper nouns and adjectives should be capitalized. Many defense-related nouns and adjectives, e.g., program and project names, weapons systems, etc., are common. Capitalization should not be used to add undue emphasis to a common noun or adjective, thereby making it "proper." If the indefinite articles "a" or "an" precede a noun or adjective, the noun or adjective is usually common. Sections B.2.1.2.1 to B.2.1.2.5 address specific areas of capitalization.

## B.2.1.2.1    Derivatives of Proper Names

Derivatives of proper names used with acquired independent common meaning or no longer identified with such names, should appear lowercase, e.g., angstrom unit, coulomb, gauss, roentgen, volt, watt, etc.

## B.2.1.2.2    Names of Military Organizations

The full names of military organizations and their shortened names should be capitalized; other substitutes, which are most often regarded as common nouns, should be capitalized only in certain specified instances to indicate preeminence or distinction. Generally, the abbreviation "U.S." should be used in lieu of "United States" when it precedes or modifies a noun. "United States" should be used as a noun. Section B.2.1.2.5 provides examples.

## B.2.1.2.3    Scientific Names

The name of a phylum, class, order, family, or genus should be capitalized. The name of a species should not be capitalized, even though derived from a proper name.

In scientific descriptions, coined terms derived from proper names should not be capitalized. Any plural formed by adding "s" to a Latin ge-

| | | | |
|---|---|---|---|
| ⊙ | Insert period | *rom.* | Roman type |
| ⋀ | Insert comma | *caps.* | Caps—used in margin |
| : | Insert colon | ≡ | Caps—used in text |
| ; | Insert semicolon | *c+sc* | Caps & small caps—used in margin |
| ? | Insert question mark | ≡ | Caps & small caps—used in text |
| ! | Insert exclamation mark | *l.c.* | Lowercase—used in margin |
| =/ | Insert hyphen | / | Used in text to show deletion or |
| ⋁ | Insert apostrophe | | substitution |
| ⋁⋁ | Insert quotation marks | ⋌ | Delete |
| ⊥N | Insert 1-en dash | ⋌ | Delete and close up |
| ⊥M | Insert 1-em dash | *w.f.* | Wrong font |
| # | Insert space | ⊂ | Close up |
| *ld>* | Insert ( ) points of space | ⊐ | Move right |
| *shill* | Insert shilling | ⊏ | Move left |
| ⋁ | Superior | ⊓ | Move up |
| ⋀ | Inferior | ⊔ | Move down |
| (/) | Parentheses | ‖ | Align vertically |
| [/] | Brackets | = | Align horizontally |
| □ | Indent 1 em | ⊐⊏ | Center horizontally |
| ⊏⊐ | Indent 2 ems | ⊔⊓ | Center vertically |
| ¶ | Paragraph | *eq.#* | Equalize space—used in margin |
| *no ¶* | No paragraph | ⋁⋁⋁ | Equalize space—used in text |
| *tr* | Transpose —used in margin* | ........ | Let it stand—used in text |
| ∩ | Transpose —used in text | *stet.* | Let it stand—used in margin |
| *sp* | Spell out | ⊗ | Letter(s) not clear |
| *ital* | Italic—used in margin | *run over* | Carry over to next line |
| ___ | Italic—used in text | *run back* | Carry back to preceding line |
| *b.f.* | Boldface—used in margin | *out, see copy* | Something omitted—see copy‡ |
| ∿∿∿ | Boldface—used in text | *9/?* | Question to author to delete‡ |
| *s.c.* | Small caps—used in margin | ⋀ | Caret—General indicator used |
| ≡≡ | Small caps—used in text | | to mark position of error. |

*In lieu of the traditional mark "tr" used to indicate letter or number transpositions, the striking out of the correct letter or numbers and the placement of the correct matter in the margin of the proof is the preferred method of indicating transposition corrections.
†Corrections involving more than two characters should be marked by striking out the entire word or number and placing the correct form in the margin. This mark should be reserved to show transposition of words.
‡The form of any query carried should be such that an answer may be given simply by crossing out the complete query if a negative decision is made or the right-hand (question mark) portion to indicate an affirmative answer.

## Figure B-1.   Editor's and Illustrator's Marks

neric name should be capitalized. In soil science, the 24 soil classifications
(Section B.2.1.2.5, list beginning on page 261) should be capitalized.

The names of the celestial bodies in the solar system, i.e., Sun, Moon
(Earth's), Mercury, Venus, Earth, Mars, Jupiter, Saturn, Uranus, Neptune,

and Pluto, should be capitalized, but the moons of (planets other than Earth) and terms like galaxy, star, and universe should be lowercase.

### B.2.1.2.4 Titles of Persons

The titles of persons should be capitalized as follows:

- (Name), President of the United States; the President, the President-elect, the Commander in Chief; similarly, the Vice President, Vice-President-elect, Vice-Presidential
- (Name), Secretary of Defense; the Secretary; similarly, the Under Secretary, the Acting Secretary, the Assistant Secretary, *but* Secretaries of the military departments
- (Name), Chairman, Joint Chiefs of Staff; Joint Chiefs of Staff; Chief of Staff, U.S. Air Force; the Chief of Staff, *but* the commanding general; general (military title standing alone not capitalized); (Name), rear admiral, U.S. Navy; the rear admiral

### B.2.1.2.5 Capitalization and Spelling Examples

The following list provides capitalization and spelling examples of terms that may be applicable to defense-related technical reports. The list is not necessarily inclusive; if a specific term is not listed, a similar example should be followed.

- A-bomb; *but* atomic bomb
- Academy
  - Air Force Academy, Coast Guard Academy, Military Academy, Naval Academy; *but* service academies
  - National Academy of Sciences; the Academy of Sciences, the academy
- Act (Federal, State, or foreign), short or popular title or with number; the act
  - Appropriation
  - Classification
- Acting, if part of capitalized title; Acting Secretary of Defense
- Adjutant General, the
- Administration, with name; capitalized standing alone if Federal unit; *but* Bush administration; administration policy
- Admiralty
  - British, etc.
  - Lord of the
- Adobe Acrobat Portable Document Format (PDF)
- Adobe Acrobat Reader

- Afghan war; Afghanistan war (undeclared by Congress)
- Agency, if part of name; capitalized standing alone if referring to U.S. unit
  - Defense Intelligence Agency; the Agency
  - *but* intelligence agencies
- agent orange
- Ages
  - atomic age
  - information age
  - missile age
  - rocket age
  - space age
- Agreement, with name; the agreement
  - General Agreement on Tariffs and Trade (GATT); the general agreement
  - Status of Forces; *but* status-of-forces agreements
- aide-de-camp
- Air Force, U.S.
  - Air National Guard
  - Base (with name); base (without name)
  - Command
  - Reserve
- Air Force One (fixed-wing Presidential aircraft)
- alliances and coalitions; *see also* North Atlantic Treaty Organization (NATO)
  - Western Powers
  - Western Union (powers); the union
- Allies, World Wars I and II and members of Western bloc (NATO); *but* one's allies, weaker allies, etc. (general)
- al-Qaida
- American Standard Code for Information Exchange (ASCII)
- appellations
  - Holocaust (World War II)
  - Third World
  - *but* war on terror; war on terrorism
- Appendix A; Appendix B: (Title); the appendix (single appendix)
- AppleShare
- AppleTalk Address Resolution Protocol (AARP)
- Arab States
- Arabic numerals
- Arm, Cavalry, Infantry, etc. (military); the arm

- Armed Forces (synonym for overall Military Establishment [U.S. and foreign]); British Armed Forces; the Armed Forces of the United States
- armed services
- Armory, 5th Regiment, etc.; the armory
- Army, U.S. or foreign, if part of name; capitalized standing alone if referring to U.S. Army
  - Active; Active Duty
  - Adjutant General, the
  - All-Volunteer
  - branches
  - Brigade, 1st, etc.; the brigade
  - Command
  - Command and General Staff College
  - Company A; A Company; the company
  - Corps
  - District of Washington (military), the district
  - Division, 1st, etc.; the division
  - Engineers (the Corps of Engineers); the Engineers; *but* Army engineer
  - Establishment
  - Field Establishment
  - Field Forces
  - 1st, 2d, etc.
  - General Staff; the Staff
  - Headquarters, 1st Brigade
  - Headquarters of the; the headquarters
  - Organized Reserves; the Reserves
  - Regular Army officer; a Regular
  - Volunteer; the Volunteers; a Volunteer
- army (general or unspecified)
  - foreign
  - mobile
  - of occupation; occupation army
- Assistant, if part of capitalized title; the Assistant Secretary of the Navy; the assistant
- Asynchronous Balanced Mode (ABM)
- Atlantic Fleet
- Ayatollah
- Balkan States
- Baltic States

- Barracks, if part of name; the barracks
- Base, (Name) Air Force Base; Air Force base (without name); the base
- battalion
- Battle, if part of name; the battle
- bill of rights, GI
- bloc
- Boolean
  - logic
  - operator
  - search
- bulletin board service (BBS)
- cache
- caliber
- caliper
- calk (spike); caulk (seal)
- Camp, if part of name; Camp Pendleton; the camp
- CD-ROM (en dash separator)
- central processing unit (CPU)
- civil defense
- client/server
- Code (in shortened title of a publication); the code
  - of Federal Regulations
  - Uniform Code of Military Justice
  - U.S.
- cold war; post-cold war
- College, if part of name; the college
  - Armed Forces Staff
  - Command and General Staff
  - National War
- COM port
- Command, capitalized with name; the command
  - Naval Sea Systems
  - North American Aerospace Defense
- Commandant, the (U.S. Coast Guard or U.S. Marine Corps only)
- Commandos, U.S., British; etc.; Commando raid; a commando
- Communist; *but* communism; communistic
- complement (complete); compliment (praise)
- Comsat
- Congressional, if part of name; Congressional Medal of Honor; *but* congressional action, committee, etc.

- Constitution, with name of country; capitalized standing alone when referring to specific national constitution (U.S. or otherwise); *but* constitutional
- Contra
- Corps, if part of name; U.S. Marine Corps; otherwise, the corps
- corpsman
- cruise missile
- D-day (day of commencement of operations)
- defense
- Defense Establishment; *but* defense community
- Department, if part of name; capitalized standing alone if referring to U.S. or international organization; Department of the Army, Department of Defense, etc.
- Depot, if part of name; the depot
- Deputy, if part of capitalized title; *but* the deputy
- Division (military), if part of name; 29th Division; the division
- Document, if part of name; the document
- Drawing, when part of title; the drawing
- dreadnought
- duffelbag
- Earth (planet); earth (soil)
- East (Communist political entity)
- electronics (noun); electronic (adjective)
- E-mail (uppercase "E" to start sentence); e-mail (lowercase within sentence)
- Engineer officer, etc. (of Engineer Corps); the Engineers
- Engineers, Chief of (Army)
- Engineers, Corps of
- Establishment, if part of name; *but* the defense establishment
- Executive Order (U.S. Presidential; with number); Executive Order 12958; EO 13292; *but* an Executive order (without number)
- Federal (synonym for U.S. or foreign government); *but* federally
- Figure (with number); Figure 1-1 (en dash separator), Figure 3, Figure A-1, etc.; *but* the figure (unnumbered)
- firewall
- flammable (*not* inflammable)
- Fleet, if part of name or referring to U.S. Navy as a whole; the fleet (general)
  - Atlantic
  - 5th, 6th, etc.
  - Marine Force

- – Naval Reserve
- – Pacific, etc. (naval)
- – U.S.
- Force(s), if part of name; the force(s)
  - – Active Duty
  - – Active Forces
  - – Air
  - – Armed (synonym for overall Military Establishment [U.S. and foreign])
  - – Army Field; the Field Forces
  - – Fleet Marine
  - – Navy Battle
  - – Navy Scouting; Reserve Force
  - – Rapid Deployment
  - – 7th Task; the task force
- Fort, if part of name; Fort Meade; the fort
- fuse (all meanings)
- fuselage
- fusillade
- gauge
- Geiger counter
- General Board (of Navy)
- General Order No. 14; General Orders No. 14; a general order
- General Schedule
- generalissimo
- Government
  - – British, etc.; the Government
  - – department, officials, -owned, publications, etc. (U.S. Government)
  - – U.S.; National; Federal; the Government
  - – *but* State government; State and local government
- government
  - – European governments
  - – Federal, State, and municipal governments
  - – military
- Government information product
- governmental
- gray (*not* grey)
- Green Berets (U.S. Army); *but* a green beret
- Guard, National
- guardsman

- guerrilla (warfare)
- Gulf war(s); first Gulf war; second Gulf war (undeclared by Congress)
- Handbook (with number); the handbook
- hangar (aircraft enclosure)
- H-bomb; *but* hydrogen bomb
- Headquarters
  - Alaska Command; the command headquarters
  - 4th Regiment Headquarters; regimental headquarters
  - 32d Division Headquarters; the division headquarters
- H-hour (hour of commencement of operations)
- home page
- homeland, the
- information
  - age
  - technology
- insurgent
- intelligence community
- interface
- Internet, Intranet; Internet service provider
- Iraq war (undeclared by Congress)
- italic
- Joint Chiefs of Staff; the Chiefs of Staff
- Kermit
- Korean war (undeclared by Congress)
- Laboratory; if part of name; Naval Research Laboratory; capitalized standing alone if referring to U.S. unit; *but* laboratory (non-U.S.)
- law, Walsh-Healey, etc.; law 176; copyright law; Ohm's law
- Letters Patent No. 378,964; *but* patent no. 378,964; letters patent
- Library
  - Army; the library
  - of Congress; the library
- listserv
- locator service
- MacTCP
- MacWais
- Marine Corps, U.S.; the corps
  - Marines (the corps); *but* marines (individuals)
  - Organized Reserve; the Reserve
  - *also* a marine; a woman marine; the women marines (individuals); soldiers, sailors, coastguardsmen, and marines

- Marine One (rotary-wing Presidential aircraft)
- M-day (mobilization day)
- Middle East
- military (general); the U.S. military; *but* Military Establishment
- Military Handbook, Specification, Standard, etc. (usually abbreviated with number); *but* a military handbook, specification, standard, etc.
- Militia, if part of name; the militia
- missiles: capitalize but do not italicize specific missiles
  - Harpoon missile
  - Scud missile
  - Standard missile; *but* Standard Missile-2
  - *but* cruise missile, surface-to-air missile
- missilry
- Moon (Earth); moon (planets other than Earth)
- Muslim
- Nation (synonym for United States); *but* a nation, French nation
- National, in conjunction with capitalized name
  - Academy of Sciences
  - Guard, Ohio, etc.; Air National; the National Guard; *but* a National Guard man; National Guardsman
  - Naval Medical Center (Bethesda, MD)
  - Science Foundation; the Foundation
  - War College
- national defense agencies
- Naval, if part of name
  - Academy
  - Base, Guam Naval; the naval base
  - District, 1st Naval
  - Establishment
  - Militia; the militia
  - Reserve Force; the force
  - Reserve officer; a Reserve officer
  - Shipyard (if preceding or following name); *but* the naval shipyard
  - Station (if preceding or following name); the station
  - Volunteer Naval Reserve
  - War College; the War College; the college
- naval (general)
  - command
  - district

- expenditures, maneuvers, officer, service, stores, etc.
- petroleum reserves; *but* Naval Petroleum Reserve No. 2; reserve no. 2
- Navy, U.S. or foreign, if part of name; capitalized standing alone if referring to U.S. Navy
  - Battle Force, the Battle Force; the force
  - Establishment; the establishment
  - Regular
  - Scouting Force; the scouting force; the force
  - Seabees (construction battalion); a Seabee
  - 7th Task Force
  - Yard; the Washington Navy Yard
- network
- North Atlantic Treaty Organization (NATO)
  - Chiefs of Staff
  - Committee of Defense Ministers
  - Council of Foreign Ministers
  - Defense Committee
  - Military Committee
  - Military Production and Supply Board
  - Mutual Defense Assistance Program
  - Pact
  - Regional Planning Group; the Group
  - Standing Group; the Group
- North Korea
- numbers, capitalized if spelled out as part of name; Air Force One; Marine One
- Office, if referring to U.S. unit; the office
  - Naval Oceanographic
  - of the Chief of Naval Operations
  - of the Secretary of Defense: Secretary's Office
- officer
  - Army
  - Marine; *but* naval and marine officers
  - Navy; Navy and Marine officers
  - Regular Army: Regular; a Regular
  - Reserve
- online
- Operation Desert Storm, Operation Iraqi Freedom, Operation Enduring Freedom, etc.; *but* Enduring Freedom operation

- Ordnance (military supplies); ord*i*nance (local or municipal regulation)
    - Corps
    - Department; the Department
    - Depot
- Organized
    - Marine Corps Reserve; Marine Reserve; the Reserve
    - Naval Militia; the Naval Militia; the militia
    - Reserve Corps; the Reserve
- Pararescue (U.S. Air Force); *but* a pararescueman
- Part (with number); Part II; the part
- permanent access service
- Persian Gulf; Persian Gulf States; Persian Gulf war (undeclared by Congress)
- President
    - of any country; the President; *but* the company president; the president of the university
    - preceding name; of the United States; the Executive; the Chief Magistrate; the Commander in Chief; the President-elect; ex-President; former President
- Presidential authority, Executive order, proclamation, etc.
- Prime Minister (of any country)
- private key
- Program, if part of name; the program
- Program Element (with number)
- Project, if part of name; the project
- Proving Ground, Aberdeen, etc.; the proving ground
- public key
- Public Law *or* P.L. (with number); a public law
- Purple Heart; purple (joint service) organization
- Rangers (U.S. Army); *but* a ranger
- README file
- reconnaissance
- Regular Army, Navy; a Regular
- Regulation (with number); the regulation
- Report, if part of name (with date or number); the report
- Reserve, if part of name; the Reserve
- Reserves; the reservist
- Roman numeral; *but* roman type
- Rule (with number); the rule

- School, if part of name; Warrant Officer Candidate School; the school
- Seabees
- SEALs (Sea, Air, Land); a SEAL
- Second World War
- Secretaries of the Army and Navy; *but* Secretaries of the military departments; secretaryship
- Secretary, head of U.S. Government organization; Secretary of the Air Force
- Section (with number); Section 2; the section
- Secure Sockets Layer
- server
- Service, if referring to U.S. unit; Selective Service System; the Service
- service
  - Army
  - naval
  - Navy
  - military
- ships, ship names and classes capitalized and italicized
  - HMS *Ark Royal*
  - *Ohio* class submarine
  - USS *Dwight D. Eisenhower* (CV 69); the *Dwight D. Eisenhower*
  - *but* a Trident submarine; a submarine tender
- soil classifications
  - Alpine Meadow
  - Bog
  - Brown
  - Chernozem (Black)
  - Chestnut
  - Desert
  - Gray-Brown Podzolic
  - Half Bog
  - Laterite
  - Pedalfer
  - Pedocal
  - Podzol
  - Prairie
  - Ramann's Brown
  - Red
  - Rendzina

- – Sierozem (Gray)
- – Solonchak
- – Solonetz
- – Soloth
- – Terra Rossa
- – Tundra
- – Wiesenboden
- – Yellow
- South Korea
- Southeast Asia
- space shuttle; the shuttle
- space station
- Special Forces (U.S. Army)
- specialist
- Specification (with number); the specification
- sputnik; *but* Sputnik I, etc.
- Standard (with number); the standard
- star wars (Strategic Defense Initiative)
- Stealth bomber, fighter, etc.
- Superintendent, if referring to head of U.S. organization; the Superintendent, U.S. Naval Academy
- supersede
- System, if referring to U.S. unit; the System
- Table (with number); Table 1-1 (en dash separator), Table 3, Table A-1, etc.; *but* the table (unnumbered)
- Taliban
- task force
- Team, if part of name; the team
- Technical Report (with number); the technical report
- Telnet
- terrorism
- The, part of formal name, capitalized: The Hague; The Netherlands; *but* the Adjutant General; the USS *John F. Kennedy*
- Title (with number)
- Tris (chemical)
- Under Secretary, if referring to officer of U.S. Government; the Under Secretary
- Uniform Code of Military Justice
- Unit, if referring to U.S. unit; the Unit
- universal military training
- University, if part of name; the university

- Vietnam war (undeclared by Congress)
- Volume (with number)
- Volunteer Naval Reserve
- War, if part of formal name and declared by Congress
  - First World War; World War I; Great War (World War I); Second World War, World War II
  - post-World War II
  - the two World Wars
- war, descriptive, undeclared by Congress, or hypothetical
  - cold, hot; post-cold war
  - Gulf
  - in Afghanistan
  - in Iraq
  - Korean
  - Persian Gulf
  - third world; world war III
  - Vietnam
  - war on terror; war on terrorism
- War College, National
- warfighter
- weapons and weapon systems
  - Aegis Combat System; the combat system
  - M-1 Abrams Main Battle Tank; the main battle tank
  - *but* air defense system, electronic warfare system
- weapons of mass destruction
- Web
  - broadcasting
  - browser
  - site
  - Webcasting
  - WebTV
- West (world political entity)
- Western
  - bloc
  - Europe(an) (political entity)
  - Powers
- white paper (report)
- Wide Area Information Server (WAIS)
- wide area network (WAN)
- WinWAIS
- woman marine, sailor, soldier, etc.

- World War I; World War II; *not* World War 1 *or* 2 *or* World War One *or* Two; World War II veteran
- x ray, x raying (no hyphen)
- Xmodem
- Ymodem
- Zmodem

### B.2.1.3    Compounding Rules

A compound word is a union of two or more words, either with or without a hyphen. Some two-word terms should not be hyphenated.

### B.2.1.3.1    Military Titles

A military title denoting a single office should not be hyphenated.
- Commander in Chief
- lieutenant commander
- major general
- Vice President
- Under Secretary
- *but* Under-Secretaryship; Vice-Presidency

### B.2.1.3.2    Scientific and Technical Terms

A hyphen should not appear in scientific terms (names of chemicals, diseases, animals, insects, or plants) used as unit modifiers if no hyphen appears in their original form.
- carbon monoxide poisoning
- methyl bromide solution
- stem rust control
- equivalent uranium content

A hyphen should appear in chemical elements used in combination with figures, except with superscript figures.
- Freon-12
- polonium-210
- uranium-235
- $U^{235}$
- $Sr^{90}$
- $_{92}U^{234}$

Hyphens and closeup punctuation should be used in chemical formulas.
- 9-nitroanthra(1,9,4,10)bis(1)oxathiazone-2,7-bisdioxide
- Cr-Ni-Mo
- 2,4-D

A hyphen should appear between the elements of technical or contrived compound units of measurement.

- candela-hour
- horsepower-hour
- light-year
- passenger-mile
- staff-hour
- work-year
- *but* kilowatthour

## B.2.1.4   Symbols, Abbreviations, and Acronyms

### B.2.1.4.1   Symbols and Abbreviations

Generally, a period should not follow a symbol or an abbreviated unit of measurement, e.g., "lb" not "lb." for pound(s). However, where the omission of a period causes possible confusion in the text, e.g., "in" for inch (not to be mistaken for the preposition "in"), a period should be placed after the abbreviation, e.g., "in." The abbreviation "in" is acceptable when combined with another abbreviation, e.g., "in-lb" for "inch-pound(s)."

The use of symbols for units of measurement, e.g., "%" for "percent," should appear only in figures and tables and equations; however, a symbol combined with an abbreviation, e.g., "72 °F," is preferred over "72 degrees Fahrenheit" or "72 deg F" in the text and in figures and tables. The symbols "'" and "''" should not be used as abbreviations for feet and inches, respectively; instead, "ft" and "in." should be used. Symbols and abbreviations should appear only with corresponding numbers, not by themselves. For example, "The temperature was measured in degrees Celsius"; not "The temperature was measured in degrees C," or "The temperature was measured in °C."

### B.2.1.4.2   Acronyms

Acronyms, as stated previously, are considered abbreviations and, like capitalization, are a major issue in defense-related technical reports because of their frequent use. In a strict sense, an acronym is an abbreviation pronounced as a word; however, the term is generally more inclusive and includes all abbreviations formed from the first letter of a series of words—known as initialisms.

Acronyms should be used for the convenience of the reader, not the author. Acronyms should be identified at their first appearance in the abstract, the executive summary, and in the main body of the technical report and should, ideally, be able to be immediately recognized and comprehended by the reader thereafter; otherwise, they hinder reading. An acronym may be defined more than once in the main body of a technical report if the situation warrants it, e.g., a lengthy main body.

The spelled-out term should appear first, followed by the acronym in parentheses. Generally, an acronym should not accompany a spelled-out term if the spelled-out term only appears once in the abstract, the executive summary, or the main body of the technical report; however, in defense-related technical reports, the acronym may be more recognizable to the reader than the spelled-out term, e.g., "IEEE" for "Institute of Electrical and Electronics Engineers," so this rule may be waived in those instances.

The definite article "the" may precede certain acronyms, e.g., "the FBI," in the text to maintain a common description of familiar terms. Other acronyms, e.g., "DoD," may or may not be preceded by "the." In the latter case, either description is acceptable; however, consistency should be maintained.

The following guidelines should be used when creating an acronym:

- All-capital letters should be used when only the first letter of each term or selected terms is used to make up the acronym.
  - MAG (*M*ilitary *A*dvisory *G*roup)
  - MIRV (*m*ultiple *i*ndependently targetable *r*eentry *v*ehicle)
  - SALT (*s*trategic *a*rms *l*imitation *t*alks; *avoid* SALT talks)
- All-capital letters should be used when the first letters of prefixes and/or suffixes are used as part of the acronym.
  - AAW (*a*nti-*a*ir *w*arfare)
  - FLIR (*f*orward-*l*ooking *i*nfra*r*ed)
- Capital and lowercase letters should be used when proper names are used in shortened form, any term of which uses more than the first letter of each term.
  - Aramco (*Ar*abian-*Am*erican Oil *Co*mpany)
  - Unprofor (*Un*ited *N*ations *Pro*tection *For*ce)
- Lowercase letters should be used in common noun combinations made up of more than the first letter of lowercased terms.
  - loran (*lo*ng-*ra*nge *n*avigation)
  - secant (*se*paration *c*ontrol of *a*ircraft by *n*onsynchronous *t*echniques)
  - sonar (*so*und *na*vigation *r*anging)
- Some terms, such as ACTION (an agency of the Government), are not acronyms but, nevertheless, should be spelled in all-capital letters.

## B.2.1.4.3    Standard Abbreviations

The number of abbreviations —specifically acronyms—used in the defense community is unlimited and increasing. Many of the new abbreviations pertain to information technology or computers. Some terms have more than one abbreviation. Table B-1 lists some of the standard abbreviations that may appear in defense-related technical reports and indicates whether the abbreviation and its spelled-out equivalent should be capitalized or appear lowercase. (Other abbreviations used herein are defined in the list of abbreviations, beginning on page 315.)

### Table B-1.    Standard Abbreviations

| Abbreviation | Term |
|:---:|:---:|
| 2,4-D | (insecticide) |
| 9/11 | 11 September 2001 |
| 24/7 | 24 hours a day, 7 days a week |
| AAM | air-to-air missile |
| AARP | AppleTalk Address Resolution Protocol |
| abbr. | abbreviation |
| ABM | anti-ballistic missile |
| abs. | abstract |
| AC | active component |
| ACDA | Arms Control and Disarmament Agency |
| A/D | analog/digital |
| ADN | advanced digital network |
| ADP | automated data processing |
| ADSL | asymmetric digital subscriber line |
| AF | audiofrequency |
| AFB | Air Force Base (with name) |
| AFRL | Air Force Research Laboratory |
| a.k.a. | also known as |
| AM | amplitude modulation |
| a.m. | (ante meridiem) before noon |
| antilog | antilogarithm |

Table B-1.   Standard Abbreviations (Continued)

| Abbreviation | Term |
| --- | --- |
| AO | area of operations |
| AOR | area of responsibility |
| APO (no periods) | Army post office |
| APPR | Army package power reactor |
| approx. | approximately |
| ARL | Army Research Laboratory |
| ARNG | Army National Guard |
| ARP | Address Resolution Protocol |
| ASCII | American Standard Code for Information Exchange |
| ASM | air-to-surface missile |
| ASME | American Society of Mechanical Engineers |
| A.S.N. | Army service number |
| ASTM | American Society for Testing and Materials |
| ATF | (Bureau of) Alcohol, Tobacco, Firearms and Explosives |
| AUS | Army of the United States |
| AUV | autonomous underwater vehicle |
| A/V | audio/visual |
| AZ | azimuth |
| AWACS | airborne warning and control system |
| AWOL | absent without leave |
| bf | boldface |
| BIOS | Basic Input/Output System |
| BIT | BInary digiT |
| BOS | battlefield operating system |
| BPS | bits per second |
| $C^2$ or C2 | command and control |

### Table B-1.   Standard Abbreviations (Continued)

| Abbreviation | Term |
|---|---|
| C$^4$ISR *or* C4ISR | command, control, communications, computers, intelligence, surveillance, and reconnaissance |
| ca. | (circa) about |
| CAD | computer-aided design |
| c. and s.c. | caps and small caps |
| CAP | Civil Air Patrol |
| CAT (scan) | computerized axial tomography |
| CBR | chemical, biological, and radiological |
| CCITT | Consultative Committee for International Telegraphy and Telephony |
| CD | compact disk |
| CD-I | Compact Disk-Interactive |
| CD-R | Compact Disk-Recordable |
| cf. | (confer) compare, or see |
| CFR | Code of Federal Regulations |
| CFR Supp. | Code of Federal Regulations Supplement |
| CG | center of gravity |
| CGI | Common Gateway Interface |
| CHAMPUS | Civilian Health and Medical Program of the Uniformed Services |
| CIA | Central Intelligence Agency |
| CIC | Counterintelligence Corps |
| CO | commanding officer |
| COA | course of action |
| con. *or* cont. | continued |
| conelrad | control of electromagnetic radiation (civil defense) |
| CONOPS | concept of operations |

### Table B-1.  Standard Abbreviations (Continued)

| Abbreviation | Term |
|---|---|
| CONUS; Conus | continental United States |
| COP | common operational picture |
| cos | cosine |
| cosh | hyperbolic cosine |
| cot | cotangent |
| coth | hyperbolic cotangent |
| COTS | commercial off-the-shelf |
| c.p. | chemically pure |
| CPR | cardiopulmonary resuscitation |
| CPU | central processing unit (computer) |
| CS | combat support |
| csc | cosecant |
| csch | hyperbolic cosecant |
| CSS | combat service support |
| CTI | computer-telephony integration |
| CY | calendar year |
| DA | Department of the Army |
| DAF | Department of the Air Force |
| DARPA | Defense Advanced Research Projects Agency |
| d.b.a. | doing business as |
| d.b.h. | diameter at breast height |
| DBMS | database management system |
| DDT | dichlorodiphenyltrichloroethane |
| DEW | distant early warning (DEW line) |
| DFT | dry film thickness |
| DHS | Department of Homeland Security |
| DIA | Defense Intelligence Agency |
| DISA | Defense Information Systems Agency |
| DMI | Desktop Management Interface |

Table B-1.   Standard Abbreviations (Continued)

| Abbreviation | Term |
| --- | --- |
| DMZ | demilitarized zone |
| DN; DON; DoN | Department of the Navy |
| DNA | deoxyribonucleic acid |
| DNS | Domain Name Service |
| do. | (ditto) the same |
| DODISS | Department of Defense Index of Specifications and Standards |
| DOF | degree(s) of freedom |
| DOJ; DoJ | Department of Justice |
| DOS | disk operating system |
| DOTMLPF | doctrine, organization, training, materiel, leadership and education, personnel, and facilities |
| dpi | dots per inch |
| DS | direct support |
| DSO | domestic support operations |
| DSSL | Document Style and Semantics Language |
| d.s.t. | daylight saving (no "s") time |
| DSWA | Defense Special Weapons Agency |
| DTP | desktop publishing |
| DVD | digital video disk |
| EBCDIC | Extended Binary Coded Decimal Interchange Code |
| e.g. | for example |
| EHF | extremely high frequency |
| EL | elevation |
| ELINT | electronics intelligence |
| e-mail | electronic mail |
| EMP | electromagnetic pulse |
| EOP | Executive Office of the President |
| EPA | Environmental Protection Agency |

**Table B-1.  Standard Abbreviations (Continued)**

| Abbreviation | Term |
|---|---|
| EPS | Encapsulated PostScript file |
| et al. | (et alii) and others |
| et seq. | (et sequentia) and the following |
| ETA | estimated time of arrival |
| etc. | (et cetera) and so forth |
| EU | European Union |
| Euratom | European Atomic Energy Community |
| Euro | currency (common) |
| Eurodollars | U.S. dollars used to finance foreign trade |
| EW | electronic warfare |
| f., ff. | and following page (pages) |
| FAA | Federal Aviation Administration |
| FAQ | frequently asked question(s) |
| f.a.s. | free alongside ship |
| FAX | facsimile |
| FBI | Federal Bureau of Investigation |
| FCC | Federal Communications Commission |
| FCS | future combat system |
| FDA | Food and Drug Administration |
| FDDI | Fiber Distributed Data Interface |
| FFIR | friendly force information requirements |
| FLOT | forward line of own troops |
| FM | frequency modulation |
| f$^o$ | folio |
| f.o.b. | free on board |
| FORCEnet | U.S. Navy enterprise network |
| FOSI | Formatting Output Specification Instance |

**Table B-1.   Standard Abbreviations (Continued)**

| Abbreviation | Term |
|---|---|
| FPO (no periods) | Fleet post office |
| FSCL | fire support coordination line |
| FTP | File Transfer Protocol |
| FWA | fraud, waste, and abuse |
| FY | fiscal year |
| GAO | General Accounting Office |
| GATT | General Agreement on Tariffs and Trade |
| GDI | Graphical Device Interface |
| GGP | Gateway-to-Gateway Protocol |
| GI | general issue; Government issue |
| GIF | Graphics Interchange Format |
| GILS | Government Information Locator Service |
| GMT | Greenwich Mean Time |
| GNP | gross national product |
| GOTS | Government off-the-shelf |
| GPO | Government Printing Office |
| GPS | global positioning system |
| gr. wt. | gross weight |
| GSA | General Services Administration |
| GUI | Graphical User Interface |
| GWOT | global war on terrorism |
| hazmat | hazardous material |
| HD | homeland defense |
| HDBK | handbook |
| HE | high explosive |
| HF | high frequency |
| HLS | homeland security |
| HMMWV | high mobility multipurpose wheeled vehicle |

### Table B-1.  Standard Abbreviations (Continued)

| Abbreviation | Term |
| --- | --- |
| HMS (not italic; no periods) | Her/His Majesty's Ship |
| HQ | headquarters |
| HS | homeland security |
| HTML | HyperText Markup Language |
| HTTP | HyperText Transfer Protocol |
| HTTPD | HyperText Transfer Protocol Daemon |
| HUMINT | human intelligence |
| HYTELNET | HyperText-browser for Telnet Accessible Sites |
| IADB | Inter-American Defense Board |
| IAEA | International Atomic Energy Agency |
| ibid. | (ibidem), in the same place |
| ICBM | intercontinental ballistic missile |
| ICMP | Internet Control Message Protocol |
| ID | Identification |
| id. | (idem), the same |
| IDA | Institute for Defense Analyses |
| IDE | integrated drive electronics |
| i.e. | (id est), that is |
| IED | improvised explosive device |
| IEEE | Institute of Electrical and Electronics Engineers |
| IF | intermediate frequency |
| Inmarsat | International Maritime Satellite |
| INS | Immigration and Naturalization Service |
| Interpol | International Criminal Police Organization |
| I/O | input/output |
| IO | information operations |
| IP | Internet Protocol |

Table B-1.    Standard Abbreviations (Continued)

| Abbreviation | Term |
| --- | --- |
| IPB | intelligence preparation of the battlespace |
| IR | infrared |
| IRBM | intermediate range ballistic missile |
| IRE | Institute of Radio Engineers |
| IRS | Internal Revenue Service |
| ISDN | Integrated Services Digital Network |
| ISP | Internet service provider |
| ISR | intelligence, surveillance, and reconnaissance |
| IT | information technology |
| ITU | International Telecommunications Union |
| JAG | Judge Advocate General |
| jato | jet-assisted takeoff |
| JCS | Joint Chiefs of Staff |
| JIT | just in time |
| JPEG | Joint Photographic Experts Group |
| Jpn. | Japan or Japanese where necessary to abbreviate |
| LABEVAL | laboratory evaluation |
| LAN | local area network |
| LAPM | Link Access Procedure for Modems |
| lat. | latitude |
| LC | Library of Congress |
| lc. | lowercase |
| LCD | liquid crystal display |
| LEO | low Earth orbit |
| LF | low frequency |
| lf. | lightface |
| liq. | liquid |

### Table B-1.   Standard Abbreviations (Continued)

| Abbreviation | Term |
|:---:|:---:|
| LOC | line of communication |
| loc. cit. | (loco citato), in the place cited |
| log | logarithm |
| long. | longitude |
| loran | long-range navigation |
| lox | liquid oxygen |
| LPG | liquefied petroleum gas |
| Ltd. | limited |
| *M* | more |
| m. | (meridies), noon |
| MAC | Military Airlift *or* Assistance Command |
| MACOM | major Army command |
| MAG | Military Advisory Group |
| MARPOL | International Convention for the Prevention of Pollution from Ships |
| MDA | Missile Defense Agency |
| MDAP | Mutual Defense Assistance Program |
| MEDEVAC | medical evacuation |
| memo | memorandum |
| METL | mission essential task list |
| MF | medium frequency; microfiche |
| MFN | most favored nation |
| MIA | missing in action (plural, MIAs) |
| MIL–HDBK | military handbook |
| MIL-STD | military standard |
| MIME | Multipurpose Internet Mail Extension |
| MIPS | millions of instructions per second |
| MIRV | multiple independently targetable reentry vehicle |

**Table B-1.    Standard Abbreviations (Continued)**

| Abbreviation | Term |
| --- | --- |
| Misc. Doc. (with number) | miscellaneous document |
| MOS | military occupational specialty |
| MP | military police |
| MPEG | Motion Pictures Experts Group |
| MRI | magnetic resonance imaging |
| M&S | modeling and simulation |
| MS, MSS | manuscript, manuscripts |
| MSC | Military Sealift Command |
| MS-DOS | Microsoft-Disk Operating System |
| m.s.l. | mean sea level |
| MTBF | mean time between failure |
| MTBR | mean time between repair |
| MTTF | mean time to failure |
| MTTR | mean time to repair |
| NA | not applicable; not available |
| NAFTA | North American Free Trade Agreement |
| NAS | National Academy of Sciences |
| NASA | National Aeronautics and Space Administration |
| NAT | network address translation (module) |
| NBC | nuclear, biological, and chemical |
| NBCC | nuclear, biological, chemical, and conventional |
| NEC | National Electrical Code |
| n.e.c. | not elsewhere classified |
| n.e.s. | not elsewhere specified |
| net wt. | net weight |
| NIH | National Institutes (plural) of Health |
| N-ISDN | Narrowband-Integrated Services Digital Network |

## Table B-1.  Standard Abbreviations (Continued)

| Abbreviation | Term |
|---|---|
| NIST | National Institute of Standards and Technology |
| n.l. | natural log or logarithm |
| NNTP | Network News Transfer Protocol |
| No., Nos. (with number[s]) | number, numbers |
| NOAA | National Oceanic and Atmospheric Administration |
| n.o.i.b.n. | not otherwise indexed by name |
| n.o.p. | not otherwise provided (for) |
| n.o.s. | not otherwise specified |
| NOVS | National Office of Vital Statistics |
| NRC | Nuclear Regulatory Commission |
| NRL | Naval Research Laboratory |
| NRO | National Reconnaissance Office |
| NS | nuclear ship |
| NSA | National Security Agency |
| NSC | National Security Council |
| NSF | National Science Foundation |
| n.s.k. | not specified by kind |
| NSN | national stock number |
| n.s.p.f. | not specifically provided for |
| OAPEC | Organization of Arab Petroleum Exporting Countries |
| OCD | Office of Civil Defense |
| OCR | optical character recognition |
| OD | officer of the day |
| OIF | Operation Iraqi Freedom |
| OJT | on-the-job training |
| OOB | order of battle |
| op. cit. | (opere citato), in the work cited |
| OPCON | operational control |

Table B-1.   Standard Abbreviations (Continued)

| Abbreviation | Term |
|---|---|
| OPEC | Organization of Petroleum Exporting Countries |
| OPEVAL | operational evaluation |
| OSD | Office of the Secretary of Defense |
| PC | personal computer |
| PDL | Page Description Language |
| PEO | Program Executive Office |
| PERL | practical extraction and report language |
| Ph | phenyl |
| PIN | personal identification number |
| PING | Packet Internet Groper |
| PKI | Public Key Infrastructure |
| p.m. | (post meridiem), after noon |
| P.O. Box | Post Office Box (with number); *but* post office box (in general sense) |
| POC | point of contact |
| POP | Point of Presence; Post Office Protocol |
| POW | prisoner of war (plural, POWs) |
| PPP | Point-to-Point Protocol |
| PPTP | Point-to-Point Tunneling Protocol |
| pro tem | (pro tempore), temporarily |
| P.S. | (post scriptum), postscript |
| PTSD | post-traumatic-stress syndrome |
| PVC | polyvinyl chloride |
| QA | quality assurance |
| QT | on the quiet |
| racon | radar beacon |
| radar | radio detection and ranging |
| RAID | redundant array of inexpensive disks |

## Table B-1.   Standard Abbreviations (Continued)

| Abbreviation | Term |
|---|---|
| RAM | Random-Access Memory |
| rato | rocket-assisted takeoff |
| RC | Reserve component |
| RF | radio frequency |
| Rh | Rhesus (blood factor) |
| RIF | reduction(s) in force; RIFed, RIFing, RIFs |
| ROE | rules of engagement |
| ROK | Republic of Korea (South) |
| ROTC | Reserve Officers' Training Corps |
| RPC | remote procedure call |
| RPG | rocket-propelled grenade |
| RTF | Rich Text Format |
| SAC | Strategic Air Command |
| SAE | Society of Automotive Engineers |
| SALT | strategic arms limitation talks |
| SAM | surface-to-air missile |
| SATCOM | satellite communications |
| SBIR | Small Business Innovation Research |
| s.c. | small caps |
| sc. | (scilicet), namely |
| s.d. | (sine die), without date |
| SDI | Strategic Defense Initiative |
| sec | secant |
| sech | hyperbolic secant |
| SGML | Standard Generalized Markup Language |
| SHF | superhigh frequency |
| SHIPEVAL | shipboard evaluation |
| shoran | short range (radio) |

## Table B-1.   Standard Abbreviations (Continued)

| Abbreviation | Term |
| :---: | :---: |
| SI | Systeme International d'Unités |
| sic | thus |
| SIGINT | signals intelligence |
| sin | sine |
| sinh | hyperbolic sine |
| SLIP | Serial Line Internet Protocol |
| SME | subject matter expert |
| SMTP | Simple Mail Transfer Protocol |
| S/N | serial number |
| SNMP | Simple Network Management Protocol |
| sonar | sound, navigation, and ranging |
| SOP | standard operating procedure |
| SOW | statement of work |
| SP | shore patrol |
| sp. gr. | specific gravity |
| SPAR | Coast Guard Women's Reserve (*semper paratus—always ready*) |
| SS | steamship |
| SSL | Secure Sockets Layer |
| SSM | surface-to-surface missile |
| SSS | Selective Service System |
| S&T | science and technology |
| Stat. | Statutes at Large |
| STD | standard |
| STP | standard temperature and pressure |
| Supt. | superintendent |
| Surg. Gen. | Surgeon General |
| SWAIS | Simple Wide Area Information System |
| SWAT | special weapons and tactics (team) |

### Table B-1.   Standard Abbreviations (Continued)

| Abbreviation | Term |
| --- | --- |
| TACON | tactical control |
| tan | tangent |
| tann | hyperbolic tangent |
| TB | tuberculosis |
| TBM | tactical ballistic missile |
| TCP/IP | Transmission Control Protocol/Internet Protocol |
| TDMA | Time Division Multiplexing Access |
| TDY | temporary duty |
| T&E | test and evaluation |
| TECHEVAL | technical evaluation |
| TESTEVAL | test evaluation |
| TIC | technical information center |
| TIFF | Tagged Image File Format |
| TM | technical manual |
| t.m. | true mean |
| TNT | trinitrotoluol |
| TQM | total quality management |
| T/R | transmit/receive |
| TT | technology transfer |
| TTP | tactics, techniques, and procedures |
| UAS | unmanned aircraft system |
| UAV | unmanned aerial vehicle |
| uc. | uppercase |
| UCAV | unmanned combat aerial vehicle |
| UCMJ | Uniform Code of Military Justice |
| UDP | User Datagram Protocol |
| UHF | ultrahigh frequency |
| UK | United Kingdom |
| UL | Unclassified Limited |

## Table B-1.    Standard Abbreviations (Continued)

| Abbreviation | Term |
|---|---|
| U.N. (with periods) | United Nations |
| UPS | uninterruptible power supply |
| URN | Uniform Resource Name/Number |
| U.S. (with periods) | United States |
| U.S.A. (with periods) | United States of America |
| USA | U.S. Army |
| USA PATRIOT (Act) | Uniting and Strengthening America by Providing Appropriate Tools Required to Intercept and Obstruct Terrorism (Act) |
| USAF | U.S. Air Force |
| USB | universal serial bus |
| U.S.C. Supp. | United States Code Supplement |
| U.S.C.A. | United States Code Annotated |
| USCG | U.S. Coast Guard |
| USIA | U.S. Information Agency |
| USMC | U.S. Marine Corps |
| USN | U.S. Navy |
| USNR | U.S. Naval Reserve |
| USNS (roman type; no periods) | United States Naval Ship |
| USS (roman type; no periods) | United States Ship |
| U.S.S.R. | Union of Soviet Socialist Republics |
| UU | Unclassified Unlimited |
| UV | ultraviolet |
| v. or vs. | (versus), against |
| VA | (Department of) Veterans Affairs; *formerly* Veterans' Administration |
| VCR | video cassette recorder |
| VDD | version description document |
| VHF | very high frequency |
| viz | (videlicet), namely |

**Table B-1.   Standard Abbreviations (Continued)**

| Abbreviation | Term |
| --- | --- |
| VLAN | virtual local area network |
| VLF | very low frequency |
| VOC | volatile organic compound |
| VRML | Virtual Reality Modeling Language |
| V/STOL | vertical and/or short take-off and landing |
| VTR | video tape recording |
| VV&A | verification, validation, and accreditation |
| WAIS | Wide Area Information Server |
| WAN | wide area network |
| wf | wrong font |
| WIPO | World Intellectual Property Organization |
| WMD | weapon(s) of mass destruction |
| w.o.p. | without pay |
| WORM | write once, read many |
| WS | weapon(s) system |
| WWW | World Wide Web |
| WYSIWYG | What you see is what you get. |
| XML | Extensible Markup Language |
| XO | executive officer |

The following Web sites are useful sources for locating and identifying abbreviations and acronyms in defense-related technical reports:

- "Abbreviations and Acronyms of the U.S. Government," www.u-lib.iupui.edu/subjectareas/gov/docs_abbrev.html
- "Acronym Finder," www.acronymfinder.com
- "Acronym Finder," www3.interscience.wiley.com/stasa/search.html
- "Joint Acronyms and Abbreviations," www.dtic.mil/doctrine/jel/dod-dict/acronym_index.html
- "Military Acronyms and Glossaries," www.ulib.iupui.edu/subjectareas/gov/military.html
- "MilitaryWords.com," www.militarywords.com.

## B.2.1.4.4    Terms of Measure

Compass directions should be abbreviated as follows:
- N.
- NE.
- E.
- SW.
- S.
- NNW.
- W.
- ESE.
- 10° N. 25° W.
- NW. by N. ¼ W.

The words latitude and longitude, followed by figures, should be abbreviated in parentheses, brackets, footnotes, sidenotes, tables, and leaderwork; the figures should always be closed up.
- lat. 52°33′05″ N.
- long. 13°21′10″ E

Latitude and longitude figures should not be broken at the end of a line; instead, the line should be spaced out. In the case of an unavoidable break at the end of a line, a hyphen should be used.

Temperature and gravity should be expressed in figures. When the degree mark is used, it should appear closed up to the capital letter, not against the figures, and a space should appear between the figures and the degree mark.
- abs, absolute
- Bé, Baumé
- °C, degree Celsius; 100 °C; *not* 100° C *or* 100°C
- °F, degree Fahrenheit; 212 °F. Without figures preceding it, °C or °F should only be used in boxheads and over figure columns in tables.
- °R, degree Rankine; 671.67 °R
- K, kelvin (no degree symbol); 273.15 K
- API, American Petroleum Institute; °API; 18 °API

Metric unit letter symbols should appear in a lowercase roman type unless the unit name has been derived from a proper name, in which case the first letter of the symbol should be capitalized (for example Pa for pascal and W for watt). The exception is the letter L for liter. The same form should be used for singular and plural. The preferred symbol for cubic centimeter is $cm^3$; cc should only be used when requested.

A space should be used between a figure and a unit symbol, except in the case of the symbols for degree, minute, and second of plane angle.
- 3 m
- 45 mm

- 25 °C
- *but* 33°15'21"

In astrophysical and similar scientific matter, magnitudes and units of time may be expressed as follows, if so written in copy:

- $5^h3^m9^s$
- $4.5^h$

## B.2.1.4.5    Units of Measurement

Table B-2 lists the standard abbreviations and symbols for units of measurement in defense-related technical reports; it is based, primarily, on comparable tables in the *United States Government Printing Office Style Manual 2000* and the *United States Government Printing Office Style Manual March 1984.*[x] The same form of the abbreviation or symbol is used for singular and plural senses, e.g., "kn" for knot or knots; "g" for acceleration of gravity, not "g's" or "gees."

**Table B-2.    Standard Abbreviations and Symbols for Units of Measurement**

| Abbreviation/Symbol | Meaning |
|:---:|:---:|
| A | ampere |
| a | are (metric 100 $m^2$); atto (prefix, $10^{-18}$) |
| aA | attoampere |
| abs | absolute (temperature and gravity) |
| ac | alternating current |
| AF | audiofrequency |
| Ah | ampere-hour |
| A/m | ampere(s) per meter |
| AM | amplitude modulation |
| asb | apostilb |
| At | ampere-turn |
| at | atmosphere, technical |
| at wt | atomic weight |

---

[x]U.S. Government Printing Office, *United States Government Printing Office Style Manual March 1984*, Washington: Superintendent of Documents, U.S. Government Printing Office, 1984.

Table B-2.    Standard Abbreviations and Symbols for Units of
Measurement (Continued)

| Abbreviation/Symbol | Meaning |
| --- | --- |
| atm | atmosphere (infrequently, As) |
| au | astronomical units |
| avdp | avoirdupois |
| B | bel |
| b | barn; bit |
| bbl | barrel |
| bbl/d | barrel(s) per day |
| Bd | baud |
| bd. ft | board foot (obsolete); use fbm |
| Bé | Baumé |
| Bev | (obsolete); see GeV |
| Bhn | Brinell hardness number |
| bhp | brake horsepower |
| bm | board measure |
| bp | boiling point |
| Btu | British thermal unit |
| bu | bushel |
| C | coulomb |
| c | centi (prefix, $10^{-2}$); cyle (radio) |
| c, ct | cent(s) |
| c•h | candela-hour |
| c/m | cycle(s) per minute |
| cal | calorie (also; $cal_{IT}$, International Table; $cal_{th}$, thermochemical |
| cc | (obsolete), use $cm^3$ |
| cd | candela (candle obsolete) |
| $cd/in^2$ | candela(s) per square inch |
| $cd/m^2$ | candela(s) per square meter |
| c.f.m. | (obsolete), use $ft^3/min$ |

### Table B-2.  Standard Abbreviations and Symbols for Units of Measurement (Continued)

| Abbreviation/Symbol | Meaning |
|---|---|
| c.f.s. | (obsolete), use $ft^3/s$ |
| cg | centigram |
| Ci | curie |
| cL | centiliter |
| cm | centimeter |
| $cm^2$ | square centimeter |
| $cm^3$ | cubic centimeter |
| cmil | circular mil |
| cP | centipoise |
| cp | candlepower |
| cSt | centistokes |
| cu ft | (obsolete), use $ft^3$ |
| cu in | (obsolete), use $in^3$ |
| cwt | hundredweight |
| D | darcy |
| d | day; deci (prefix, $10^{-1}$); pence |
| da | deka (prefix, 10) |
| dag | dekagram |
| daL | dekaliter |
| dam | dekameter |
| $dam^2$ | square dekameter |
| $dam^3$ | cubic dekameter |
| dB | decibel |
| dBu | decibel unit |
| dc | direct current |
| deg | degree |
| dg | decigram |
| dL | deciliter |
| dm | data mile; decimeter |

**Table B-2.    Standard Abbreviations and Symbols for Units of Measurement (Continued)**

| Abbreviation/Symbol | Meaning |
| --- | --- |
| $dm^2$ | square decimeter |
| $dm^3$ | cubic decimeter |
| dol | dollar |
| doz | dozen |
| dr | dram |
| dwt | deadweight tons; pennyweight |
| dyn | dyne |
| E | exa (prefix, $10^{18}$) |
| emf | electromotive force |
| emu | electromagnetic unit |
| erg | erg |
| esu | electrostatic unit |
| eV | electronvolt |
| F | farad |
| f | femto (prefix, $10^{-15}$) |
| fbm | board foot; board foot measure |
| fc | footcandle |
| fL | footlambert |
| FM | frequency modulation |
| fm | femtometer |
| ft | foot |
| ft•lb | foot-pound |
| ft•lbf | foot pound-force |
| ft/min | foot/feet per minute |
| ft-pdl | foot poundal |
| ft/s | foot/feet per second |
| $ft/s^2$ | foot/feet per second squared |
| $ft/s^3$ | foot/feet per second cubed |
| $ft^2$ | square foot |

**Table B-2.   Standard Abbreviations and Symbols for Units of Measurement (Continued)**

| Abbreviation/Symbol | Meaning |
|---|---|
| ft$^2$/min | square foot/feet per minute |
| ft$^2$/s | square foot/feet per second |
| ft$^3$ | cubic foot |
| ft$^3$/min | cubic foot/feet per minute |
| ft$^3$/s | cubic foot/feet per second |
| ftH$_2$O | conventional foot of water |
| G | gauss; giga (prefix, 10$^9$) |
| g | gram; acceleration of gravity |
| g/cm$^3$ | gram(s) per cubic centimeter |
| Gal | gal cm/s$^2$ |
| gal | gallon |
| gal/min | gallon(s) per minute |
| gal/s | gallon(s) per second |
| GB | gigabyte |
| Gb | gilbert |
| GeV | gigaelectronvolt |
| GHz | gigahertz (gigacycle[s] per second) |
| gill | gill (not abbreviated) |
| gr | grain; gross |
| H | henry |
| h | hecto (prefix, 10$^2$); hour |
| ha | hectare |
| HF | high frequency |
| hg | hectogram |
| hL | hectoliter |
| hm | hectometer |
| hm$^2$ | square hectometer |
| hm$^3$ | cubic hectometer |
| hp | horsepower |

Table B-2.   Standard Abbreviations and Symbols for Units of Measurement (Continued)

| Abbreviation/Symbol | Meaning |
| --- | --- |
| hph | horsepower-hour |
| Hz | hertz (cycle[s] per second) |
| id | inside diameter |
| ihp | indicated horsepower |
| in, in. | inch (place period after in text) |
| in/h | inch(es) per hour |
| in-lb | inch-pound |
| in/s | inch(es) per second |
| $in^2$ | square inch |
| $in^3$ | cubic inch |
| inHg | conventional inch of mercury |
| $inH_2O$ | conventional inch of water |
| J | joule |
| J/K | joule(s) per kelvin |
| K | kayser; kelvin (use without degree symbol) |
| k | kilo (prefix, 1000 or $10^3$) |
| kc | kilocycle; *see also* kHz (kilohertz), kilocycle(s) per second |
| kcal | kilocalory |
| keV | kiloelectronvolt |
| kG | kilogauss |
| kg | kilogram |
| kgf | kilogram-force |
| kHz | kilohertz (kilocycle[s] per second) |
| kL | kiloliter |
| klbf | kilopound-force |
| km | kilometer |
| km/h | kilometer(s) per hour |
| $km^2$ | square kilometer |

**Table B-2.   Standard Abbreviations and Symbols for Units of Measurement (Continued)**

| Abbreviation/Symbol | Meaning |
|---|---|
| km$^3$ | cubic kilometer |
| kn | knot (speed) |
| k$\Omega$ | kilohm |
| kt | kiloton; carat |
| kV | kilovolt |
| kVA | kilovoltampere |
| kvar | kilovar |
| kW | kilowatt |
| kWh | kilowatthour |
| L | lambert; liter |
| l/m | line(s) per minute |
| l/s | line(s) per second |
| lb | pound |
| lb ap | apothecary pound |
| lb avdp | avoirdupois pound |
| lb/ft | pound(s) per foot |
| lb/ft$^2$ | pound(s) per square foot |
| lb/ft$^3$ | pound(s) per cubic foot |
| lbf | pound-force |
| lbf/ft | pound-force foot |
| lbf/ft$^2$ | pound-force(s) per square foot |
| lbf/ft$^3$ | pound-force(s) per cubic foot |
| lbf/in$^2$ | pound-force(s) per square inch (*see also* psi) |
| lct | long calcined ton |
| ldt | long dry ton |
| LF | low frequency |
| lin ft | linear foot |
| lm | lumen |

Table B-2.    Standard Abbreviations and Symbols for Units of
Measurement (Continued)

| Abbreviation/Symbol | Meaning |
|---|---|
| lm/ft$^2$ | lumen(s) per square foot |
| lm/m$^2$ | lumen(s) per square meter |
| lm•s | lumen second |
| lm/W | lumen(s) per watt |
| lx | lux |
| M | mega (prefix, 1 million or 10$^6$) |
| m | meter; milli (prefix, 10$^{-3}$) |
| M$_1$ | monetary aggregate |
| m$^2$ | square meter |
| m$^3$ | cubic meter |
| mA | milliampere |
| MB | megabyte |
| mbar | millibar |
| Mc | megacycle; *see also* MHz (megahertz), megacycle(s) per second |
| mc | millicycle; *see also* mHz (millihertz), millicycle(s) per second |
| mcg | microgram (obsolete, use μg) |
| mD | millidarcy |
| meq | milliquivalent |
| MeV | megaelectronvolts |
| mF | millifared |
| mG | milligauss |
| mg | milligram |
| Mgal/d | million gallons per day |
| mH | millihenry |
| mho | mho (obsolete, use S, siemens) |
| MHz | megahertz |
| mHz | millihertz |

### Table B-2.   Standard Abbreviations and Symbols for Units of Measurement (Continued)

| Abbreviation/Symbol | Meaning |
|---|---|
| mi | mile (statute) |
| mi/gal | mile(s) per gallon |
| mi/h | mile(s) per hour |
| $mi^2$ | square mile |
| mil | mil |
| min | minute (time) |
| mL | milliliter |
| mm | millimeter |
| $mm^2$ | square millimeter |
| $mm^3$ | cubic millimeter |
| mmHg | conventional millimeter of mercury |
| mo | month |
| mol | mole (unit of substance) |
| mol wt | molecular weight |
| mp | melting point |
| ms | millisecond |
| Mt | megaton |
| mV | millivolt |
| MW | megawatt |
| mW | milliwatt |
| mWd/t | megawatt-day(s) per ton |
| Mx | maxwell |
| mμ | (obsolete); see nm, nanometer |
| MΩ | megohm |
| N | newton |
| n | nano (prefix, $10^{-9}$) |
| N•m | newton meter |
| $N/m^2$ | newton(s) per square meter |
| $N•s/m^2$ | newton second(s) per square meter |

Table B-2.    Standard Abbreviations and Symbols for Units of
Measurement (Continued)

| Abbreviation/Symbol | Meaning |
|---|---|
| nA | nanoampere |
| nF | nanofarad |
| nm | nanometer (millimicron, obsolete) |
| nmi | nautical mile |
| Np | neper |
| ns | nanosecond |
| nt | nit |
| od | outside diameter |
| Oe | oersted (use of A/m, ampere[s] per meter, preferred) |
| oz | ounce (avoirdupois) |
| P | peta (prefix, $10^{15}$); poise |
| p | pico (prefix, $10^{-12}$) |
| p/m | part(s) per million |
| Pa | pascal |
| pA | picoampere |
| pct | percent |
| pdl | poundal |
| pF | picofarad (micromicrofarad, obsolete); water-holding energy |
| pH | hydrogen-ion concentration |
| ph | phot; phase |
| pk | peck |
| ps | picosecond |
| psi | pound(s) per square inch (*see also* $lbf/in^2$) |
| psig | pound(s) per square inch gauge |
| pt | pint |
| pW | picowatt |
| quad | quadrillion ($10^{15}$) |

**Table B-2.   Standard Abbreviations and Symbols for Units of Measurement (Continued)**

| Abbreviation/Symbol | Meaning |
|---|---|
| qt | quart |
| R | roentgen |
| r/min | revolution(s) per minute |
| r/s | revolution(s) per second |
| rad | radian |
| rd | rad |
| rem | roentgen, equivalent man |
| RH | relative humidity |
| rms | root mean square |
| S | siemens |
| s | second (time); shilling |
| s•ft | second-foot |
| sb | stilb |
| scp | spherical candlepower |
| shp | shaft horsepower |
| slug | slug |
| sr | steradian |
| sSf | standard saybolt fural |
| sSu | standard saybolt universal |
| stdft$^3$ | standard cubic foot |
| Sus | saybolt universal second |
| T | tera (prefix, $10^{12}$); tesla |
| t | tonne (metric ton) |
| tbsp | tablespoonful |
| Tft$^3$ | trillion cubic feet |
| thm | therm |
| ton | ton (not abbreviated) |
| tsp | teaspoonful |
| Twad | twaddell |

Table B-2.    Standard Abbreviations and Symbols for Units of
Measurement (Continued)

| Abbreviation/Symbol | Meaning |
|---|---|
| u | (unified) atomic mass unit |
| UHF | ultrahigh frequency |
| V | volt |
| V/m | volt(s) per meter |
| VA | voltampere |
| var | var |
| W | watt |
| W/(m•K) | watt(s) per steradian square meter |
| W/sr | watt(s) per steradian |
| W/(sr•m$^2$) | watt(s) per steradian square meter |
| Wb | weber |
| Wh | watthour |
| wk | week |
| $x$ | unknown quantity (italic) |
| yd | yard |
| yd$^2$ | square yard |
| yd$^3$ | cubic yard |
| yr | year |
| $\mu$ | micro (prefix, $10^{-6}$); micron (name micron obsolete); use $\mu$m, micrometer |
| $\mu$A | microampere |
| $\mu$bar | microbar |
| $\mu$F | microfarad |
| $\mu$g | microgram |
| $\mu$H | microhenry |
| $\mu$in | microinch |
| $\mu$m | micrometer |
| $\mu$m$^2$ | square micrometer |
| $\mu$m$^3$ | cubic micrometer |

Table B-2.   Standard Abbreviations and Symbols for Units of
Measurement (Continued)

| Abbreviation/Symbol | Meaning |
|---|---|
| μmho | micromho (obsolete, use μS, microsiemens) |
| μs | microsecond |
| μμ | micromicron (use of compound prefixes obsolete; use pm, picometer) |
| μμf | micromicrofarad (use of compound prefixes obsolete; use pF) |
| μV | microvolt |
| μW | microwatt |
| Å | angstrom |
| ¢ | cent(s) |
| ° | degree(s) |
| °C | degree(s) Celsius |
| °F | degree(s) Fahrenheit |
| °R | degree Rankine; degree Réaumur |
| $ | dollar(s) |
| % | percent |

## B.2.1.4.6    Chemical Names and Abbreviations

Table B-3 lists the chemical names and abbreviations approved by the International Union of Pure and Applied Chemistry. The abbreviations should be used in figures and tables and equations; however, the names should be used in the text. The abbreviations should appear in roman type without periods.

Table B-3.   Chemical Names and Abbreviations

| Name | Abbreviation | Name | Abbreviation |
|---|---|---|---|
| actinium | Ac | mercury | Hg |
| aluminum | Al | molybdenum | Mo |
| americium | Am | neodymium | Nd |
| antimony | Sb | neon | Ne |

## Table B-3.   Chemical Names and Abbreviations (Continued)

| Name | Abbreviation | Name | Abbreviation |
|---|---|---|---|
| argon | Ar | neptunium | Np |
| arsenic | As | nickel | Ni |
| astatine | At | niobium | Nb |
| barium | Ba | nitrogen | N |
| berkelium | Bk | nobelium | No |
| beryllium | Be | osmium | Os |
| bismuth | Bi | oxygen | O |
| bohrium | Bh | palladium | Pd |
| boron | B | phosphorus | P |
| bromine | Br | platinum | Pt |
| cadmium | Cd | plutonium | Pu |
| calcium | Ca | polonium | Po |
| californium | Cf | potassium | K |
| carbon | C | praseodymium | Pr |
| cerium | Ce | promethium | Pm |
| cesium | Cs | protactinium | Pa |
| chlorine | Cl | radium | Ra |
| chromium | Cr | radon | Rn |
| cobalt | Co | rhenium | Re |
| copper | Cu | rhodium | Rh |
| curium | Cm | rubidium | Rb |
| dubnium | Db | ruthenium | Ru |
| dysprosium | Dy | rutherfordium | Rf |
| einsteinium | Es | samarium | Sm |
| erbium | Er | scandium | Sc |
| europium | Eu | seaborgium | Sg |
| fermium | Fm | selenium | Se |
| fluorine | F | silicon | Si |
| francium | Fr | silver | Ag |
| gadolinium | Gd | sodium | Na |

**Table B-3.   Chemical Names and Abbreviations (Continued)**

| Name | Abbreviation | Name | Abbreviation |
|---|---|---|---|
| gallium | Ga | strontium | Sr |
| germanium | Ge | sulfur | S |
| gold | Au | tantalum | Ta |
| hafnium | Hf | technetium | Tc |
| hassium | Hs | tellurium | Te |
| helium | He | terbium | Tb |
| holmium | Ho | thallium | Tl |
| hydrogen | H | thorium | Th |
| indium | In | thulium | Tm |
| iodine | I | tin | Sn |
| iridium | Ir | titanium | Ti |
| iron | Fe | tungsten | W |
| krypton | Kr | ununbium | Uub |
| lanthanum | La | ununnilium | Uun |
| lawrencium | Lr | unununium | Uuu |
| lead | Pb | uranium | U |
| lithium | Li | vanadium | V |
| lutetium | Lu | xenon | Xe |
| magnesium | Mg | ytterbium | Yb |
| manganese | Mn | yttrium | Y |
| meitnerium | Mt | zinc | Zn |
| mendelevium | Md | zirconium | Zr |

## B.2.1.5    Italic

## B.2.1.5.1    Names of Aircraft, Vessels, and Spacecraft

The names of aircraft, vessels (including ship class), and spacecraft should appear in capital and lowercase letters and italic type unless otherwise indicated. Missiles, rockets, guns and rifles, and weapons systems should appear in capital and lowercase letters and roman type.

- F/A-18 *Hornet* strike fighter; *but* Air Force One; Marine One (not single, specific aircraft)

- UH-60 *Black Hawk* helicopter
- USS *John F. Kennedy* (CV 67)
  (U.S. Navy ship names should be preceded by "the USS" or "the USNS" and followed by their hull classification symbol and number in parentheses at their first mention in an abstract, executive summary, and text of a defense-related technical report. [No hyphen or en dash should appear in the hull classification symbol and number.] Thereafter, the ship may be referred to, simply, by its name, e.g., "the *Dwight D. Eisenhower*"; however, the ship should not be referred to only by its hull classification symbol and number, e.g., "the CV 69," or by shortened or informal names like "the Eisenhower" or "the Ike.")
- HMS *Ark Royal*
- *Ohio* class submarine; *but* a Trident submarine
- *Discovery* space shuttle; *Discovery*'s (roman type "s") payload doors
- MiG (lowercase "i"); MiG-23 *Flogger*
- Standard Missile-3; *but* a Standard missile
- Titan II Gemini Launch Vehicle; *but* a Titan rocket
- Gatling gun
- M-16 rifle
- Aegis Combat System; *but* an Aegis cruiser

## B.2.1.5.2    Scientific Names

The scientific names of genera, subgenera, species, and subspecies (varieties) should be italicized but should appear in roman type in italic matter; the names of groups of higher rank than genera (phyla, classes, orders, families, tribes, etc.) should appear in roman type.

## B.2.1.5.3    Letter Designations

Letter designations in mathematical and scientific matter, except chemical symbols, should be italicized.

## B.2.1.5.4    Numerals

Most rules for the use of numerals are based on the general principle that the reader comprehends numerals more readily than numerical word expressions, particularly in scientific and technical matter.

As a rule, numbers less than 10 should be spelled out within a sentence. A figure should be used for a single number of 10 or more, with the exception of the first word of a sentence. Related numbers appearing at the beginning of a sentence separated by no more than three words should be spelled out like the first word. When 2 or more numbers appear in a sentence and 1 of them is 10 or larger, figures should be used for each number.

A specific unit of measurement, time, or money (actual or implied) should be expressed in figures; however, approximations may appear as word expressions (in text). A zero should be supplied before a decimal point if there is no whole unit, and zeros should be omitted after a decimal point unless they indicate exact measurement. The unit of measurement is plural if it exceeds 1, including fractions or percentages of 2; otherwise, it is singular.

In some instances, commas should be used as separators in numbers containing four or more figures. However, the following four-figure numbers should not contain commas:

- serial numbers
- common and decimal fractions
- astronomical and military time
- kilocycles and meters pertaining to radio
- four-figure years
- numbers of not more than four figures standing alone or part of a group of numbers containing four or less figures.

Thin spaces should be used to separate groups of three figures in a decimal fraction.

The following list provides examples:

- $4^h30^m$ or $4.5^h$, in scientific work, if so written in copy
- 0025, 2359 (astronomical and military time)
- 30 June 2005; 5-7 July 2005; 6, 7 August 2005 (*not* 6-7 August 2005); 30 September 2005 to 4 October 2005 (complete dates)
- 0.25 in.; 1.00 ft (exact measurement); 1.25 yd; *but* .30 caliber (meaning 0.30 in., bore of small arms); 30 calibers
- sp. gr. 0.9547
- gauge height 10.0 ft (exact measurement)
- 123,456,789 *and* 1,234 if part of group of numbers containing five or more figures; *but* 1234 if standing alone or if part of group of numbers containing four or less figures
- 0.123 456 789 *and* 0.1 234 (thin space separating figures) if part of group of decimals containing five or more figures; *but* 0.1234 if standing alone or if part of group of decimal fractions containing four or less figures
- 8° (figures and tables and equations); 8 deg (text)
- 47 °F; *but* 150 million degrees Fahrenheit
- 60 Hz; *but* "expressed in hertz"
- multiplied by 3; divided by 6; square root of 9; a factor of 2

- 10 yd (specific, in text); ≈ 491 °R (figures, tables, and equations); *but* about twenty feet (approximate, in text [if preferred])
- 8 by 12 in.; *not* 8x12 in.
- 4- to 6-ft gap (space after "4-")
- fourfold
- two dozen
- two orders of magnitude
- six degrees of freedom; *but* 6 DOF
- zero nautical miles; *but* 0 to 12 nmi
- 1 to 4; 1-3-5; 1:62,500
- 6 h 8 min 20 s (no commas)
- 3 fiscal years; third fiscal year; 1 calendar year
- three decades
- three quarters (9 months)
- statistics of any one year
- in a year or two
- one-half hour
- two-thirds of a mile
- the eleventh hour
- 5-day week; *not* 5-d week
- 10-year-old building
- ½-inch pipe; *not* ½-in. pipe
- 21st-century technology
- $3 billion budget
- 1st Army; 1st Cavalry Division; 2d (no "n") Brigade; 323d (no "r") Fighter Wing; 12th Regiment; 9th Naval District; 7th Fleet; 7th Air Force; 7th Task Force; *but* XII Corps (Army usage)

In chemical formulas, full-sized figures should be used before the symbol or group of symbols to which they relate and inferior figures should be used after the symbol, e.g., $6PbS \cdot (Ag,Cu)_2S \cdot 2As_2S_3O_4$.

A spelled-out number should not be repeated in figures in parentheses and vice versa.

Numbers of less than 100 preceding a compound modifier containing a figure should be spelled out, e.g., twelve 6-inch guns; *but* 150 6-inch-square samples.

Indefinite expressions should be spelled out, except with dates.

- twelvefold; hundredfold; twentyfold to thirtyfold; *but* 250-fold; 2.5-fold; 41-fold
- 1980s; mid-1990s (no apostrophe); *but* turbulent sixties

For typographic appearance and easy grasp of large numbers beginning with million, the word million or billion should be used.

- $12 million; *not* $12,000,000; *also* $10 to $20 million
- $1,270,000; *not* $1.27 million
- $500,000 to $1 million
- 10 or 20 million; between 10 and 20 million
- 300,000; *not* 300 thousand
- 1½ million; not a million and a half

The prefix "k" (meaning 1000) should not be combined with numbers, e.g., 50,000 ft, *not* 50k ft.

Mixed fractions should be expressed in figures; however, fractions standing alone or if followed by "of a" or "of an" should, generally, be spelled out.

- three-fourths of an inch; *not* ¾ inch *or* ¾ of an inch
- one-half inch
- seven-tenths of 1 percent
- three-quarters of an inch
- half an inch
- a quarter of an inch
- one-hundredth
- two one-hundredths
- one-thousandth
- ½-inch pipe
- 2½ times

A comma should not be used in any part of a built-up fraction of four or more figures or in decimals.

## B.2.2   The Chicago Manual of Style

General questions regarding style not addressed by the *United States Government Printing Office Style Manual* may be addressed by *The Chicago Manual of Style,*[y] which may be used as a secondary source on matters of style in defense-related technical reports; however, should there be a discrepancy between the *United States Government Printing Office Style Manual* and *The Chicago Manual of Style*, the *United States Government Printing Office Style Manual* takes precedence. *The Chicago Manual of Style* is referenced in the *United States Government Printing Office Style Manual.*

*The Chicago Manual of Style* is available for sale at many retail book stores and on the Internet.

---

[y]The University of Chicago Press, *The Chicago Manual of Style: The Essential Guide for Writers, Editors, and Publishers (15th Edition)*, Chicago: The University of Chicago Press, 2003.

## B.2.3    Other Style Manuals

Other recognized, generally accepted style manuals, e.g., *IEEE Standards Style Manual*[z] and *The ACS Style Guide*,[aa] may occasionally be used to deal with a particular subject not addressed by the *United States Government Printing Office Style Manual* or *The Chicago Manual of Style*. However, the use of style manuals other than the *United States Government Printing Office Style Manual* or *The Chicago Manual of Style* should be used judiciously to avoid inconsistency.

---

[z]The Institute of Electrical and Electronics Engineers, Inc., *IEEE Standards Style Manual*, http://standards.ieee.org/guides/index.html, Revised April 2002.

[aa]Dodd, J.S., Ed., *The ACS Style Guide: A Manual for Authors and Editors*, Washington: The American Chemical Society, 1997.

# *Bibliography*

Appendix A of ANSI/NISO Z39.18-2005[ab] is a selected annotated bibliography and is divided into the following seven categories:
- Writing, usage, style, grammar, and English language dictionaries
- Style manuals and guides
- Specialized dictionaries, encyclopedias, and handbooks
- Technical writing materials
- Standards and symbols
- Library reference materials
- Graphic arts
- Typography and publication design.

The following bibliography is divided into two categories and lists publications not included in Appendix A of ANSI/NISO Z39.18-2005 or lists a later edition of the same publication.

## Reference Literature

*A Dictionary of Computing*, 5th ed.,Oxford, NY: Oxford University Press, Inc., 2004.

---

[ab]National Information Standards Organization, "Scientific and Technical Reports – Preparation, Presentation, and Preservation," ANSI/NISO Z39.18-2005, Bethesda, MD: NISO, 2005.

*A Dictionary of Science*, Oxford, NY: Oxford University Press, Inc., 2003.

Archambault, A., and J-C. Corbell, Eds., *The Macmillan Visual Dictionary: Unabridged Compact Edition*, New York: Macmillan Publishing Co., 1995.

Bonk, M.R., Ed., *Acronyms, Initialisms & Abbreviations Dictionary 1997*, 21st ed., Detroit, MI: Gale Research, 1996.

Bonk, M.R., Ed., *Reverse Acronyms, Initialisms & Abbreviations Dictionary 1997*, 21st ed., Detroit, MI: Gale Research, 1996.

Bothamley, J., *Dictionary of Theories*, Canton, MI: Visible Ink Press, LLC, 2002.

Considine, G.D., Ed., *Van Nostrand's Scientific Encyclopedia*, 9th ed., New York: Van Nostrand Reinhold, 2002.

Cutler, D.W., and T.J. Cutler, *Dictionary of Naval Abbreviations*, 4th ed., Annapolis, MD: U.S. Naval Institute, 2005.

Cutler, D.W., and T.J. Cutler, *Dictionary of Naval Terms*, 6th ed., Annapolis, MD: U.S. Naval Institute, 2005.

IEEE 100, *The Authoritative Dictionary of IEEE Standards Terms*, 7th ed., New York: IEEE Standards Information Network (SIN), 2000.

Mechtly, E.A., *The International System of Units: Physical Constants and Conversion Factors*, Washington: NASA, 1973.

*Merriam-Webster's Manual for Writers and Editors*, Springfield, MA: Merriam-Webster, Inc., 1998.

Ocran, E.B., *Ocran's Acronyms: A Dictionary of Abbreviations and Acronyms Used in Scientific and Technical Writing*, Boston: Routledge & K. Paul, 1978.

## Technical Communication Literature

Alley, M., *The Craft of Scientific Writing*, New York: Springer, 1996.

Alred, G.J., C. T. Brusaw, and W.E. Oliu, *The Technical Writer's Companion*, Boston: Bedford/St. Martin's, 2002.

Andrews, D.C., *Technical Communication in the Global Community*, Upper Saddle River, NJ: Prentice Hall, 2001.

Baake, K., *Metaphor and Knowledge; The Challenges of Writing Science*, Albany, NY: State University of New York Press, 2003.

Barrass, R., *Scientists Must Write: A Guide to Better Writing for Scientists, Engineers and Students*, New York: John Wiley & Sons, Inc., 1978.

Barsha, J., et al., *The Technical Report: Its Preparation, Processing, and Use in Industry and Government*, New York: Reinhold Publishing Corp., 1954.

Barzun, J., and H.F. Graff, *The Modern Researcher*, New York: Harcourt Brace Jovanovich, 1977.

Beach, D.P., and T.K.E. Alvager, *Handbook for Scientific and Technical Research*, Englewood Cliffs, NJ: Prentice Hall, 1992.

Beer, D.F., Ed., *Writing and Speaking in the Technology Professions: A Practical Guide*, New York: IEEE Press, 2003.

Biagioli, M., and P. Galison, Eds., *Scientific Authorship: Credit and Intellectual Property in Science*, New York: Routledge, 2003.

Blake, G., and R.W. Bly, *The Elements of Technical Writing*, New York: Macmillan Publishing Co., 1993.

Blickle, M.D., *Reports for Science and Industry*, New York: Holt, 1958.

Blicq, R.S., *Technically—Write! Communication for the Technical Man*, Englewood Cliffs, NJ: Prentice-Hall, 1972.

Blicq, R.S., *Writing Reports To Get Results,* New York: IEEE Press, 1987.

Blum, D., and M. Knudson, *A Field Guide for Science Writers*, New York: Oxford University Press, 1997.

Bly, R.W., and G. Blake, *Technical Writing: Structure, Standards, and Style*, New York: McGraw-Hill, 1982.

Booth, V., *Communicating in Science: Writing and Speaking*, New York: Cambridge University Press, 1984.

Booth, W.C., G.G. Colomb, and J.M. Williams, *The Craft of Research*, Chicago: University of Chicago Press, 2003.

Brinegar, B.C., and C.B. Skates, *Technical Writing: A Guide with Models*, Glenview, IL: Scott, Foresman, 1983.

Budinski, K.G., *Engineers' Guide to Technical Writing*, Materials Park, OH: ASM International, 2001.

Burnett, R.E., *Technical Communication*, Fort Worth, TX: Harcourt College Publishers, 2001.

Chandler, H.E., *Technical Writer's Handbook*, Metals Park, OH: American Society for Metals, 1983.

Clarke, E., and V. Root, *Your Future in Technical and Science Writing*, New York: R. Rosen Press, 1972.

*Contributors' Handbook to the Defense Technical Information Center*, Alexandria, VA: DTIC, 1990.

Crouch, W.G., *A Guide to Technical Writing*, New York: Ronald Press Co., 1948.

Damerst, W.A., and A.H. Bell, *Clear Technical Communication: A Process Approach*, San Diego: Harcourt Brace Jovanovich, 1990.

Davis, D.S., *Elements of Engineering Reports*, New York: Chemical Publishing Co., 1963.

Ehrlich, E., and D. Murphy, *The Art of Technical Writing: A Manual for Scientists, Engineers, and Students*, New York: Crowell, 1964.

Eisenberg, A., *Effective Technical Communication*, New York: McGraw-Hill, 1982.

Fearing, B.E., and W.K. Sparrow, Eds., *Technical Writing: Theory and Practice*, New York: Modern Language Association of America, 1989.

Flaherty, S.M., *Technical and Business Writing: A Reader-Friendly Approach*, Englewood Cliffs, NJ: Prentice Hall, 1990.

Foster, J., *Science Writer's Guide*, New York: Columbia University Press, 1963.

Gillman, L., *Writing Mathematics Well: A Manual for Authors*, Washington: Mathematical Association of America, 1987.

Greene, M., and J.G. Ripley, *Writing by Design: A Handbook for Technical Professionals*, Englewood Cliffs, NJ: Prentice Hall, 1993.

Haines, R.W., *Roger Haines on Report Writing: A Guide for Engineers*, Blue Ridge Summit, PA: TAB Professional and Reference Books, 1990.

Hammond, E.R., *Informative Writing*, New York: McGraw-Hill, Inc., 1985.

Hancock, E., *Ideas into Words: Mastering the Craft of Science Writing*, Baltimore: Johns Hopkins University Press, 2003.

Hancock, E., *The Guide to Science Writing*, Baltimore: Johns Hopkins University Press, 2003.

Haramundanis, K., *The Art of Technical Documentation*, Maynard, MA: Digital Press, 1992.

Hays, R., *Principles of Technical Writing*, Reading, MA: Addison-Wesley, 1965.

Higham, N.J., *Handbook of Writing for the Mathematical Sciences*, Philadelphia: Society for Industrial and Applied Mathematics, 1993.

Hirsch, H.L., *Essential Communication Strategies for Scientists, Engineers, and Technology Professionals*, Piscataway, NJ: IEEE Press, 2003.

Hirsch, H.L., *The Essence of Technical Communication for Engineers: Writing, Presentation, and Meeting Skills*, New York: IEEE Press, 2000.

Hoover, H., *Essentials for the Scientific and Technical Writer*, New York: Dover Publications, 1980.

Ingre, D., *Technical Writing: Essentials for the Successful Professional*, Mason, OH: Thomson, 2003.

Jones, W.P., *Writing Scientific Papers and Reports*, Dubuque, IA: W.C. Brown Co., 1981.

Jordan, S., J.M. Kleinman, and H.L. Shimberg, Eds., *Handbook of Technical Writing Practices*, New York: Wiley-Interscience, 1971.

Katzoff, S., *Clarity in Technical Reporting*, Washington: NASA, Scientific and Technical Information Division, 1964.

King, L.S., *Why Not Say It Clearly: A Guide to Scientific Writing*, Boston: Little, Brown, 1978.

Knuth, D.E., T. Larrabee, and P.M. Roberts, *Mathematical Writing*, Washington: Mathematical Association of America, 1989.

Lannon, J.M., *Technical Writing*, Boston: Little, Brown, 1979.

Lee, M., et al., *The Handbook of Technical Writing: Form and Style*, San Diego: Harcourt Brace Jovanovich, 1990.

Locke, D., *Science as Writing*, New Haven, CT: Yale University Press, 1992.

Lutz, J.A., and C.G. Storms, Eds., *The Practice of Technical and Scientific Communication: Writing in Professional Contexts*, Stamford, CT: Ablex Publishing Corp., 1998.

Markel, M., *Writing in the Technical Fields: A Step-by-Step Guide for Engineers, Scientists, and Technicians*, Piscataway, NJ: IEEE Press, 1994.

Markel, M.H., *Technical Writing: Situations and Strategies*, New York: St. Martin's Press, 1988.

McMurrey, D.A., *Processes in Technical Writing*, New York: Macmillan, 1988.

Meredith, P., *Instruments of Communication*, Oxford, NY: Pergamon Press, 1966.

Miles, T.H., *Critical Thinking and Writing for Science and Technology*, San Diego: Harcourt Brace Jovanovich, 1990.

Moriarty, M.F., *Writing Science Through Critical Thinking*, Sudbury, MA: Jones & Bartlett Publishers, 1997.

Nelson, J.R., *Writing the Technical Report*, New York: McGraw-Hill Book Company, Inc., 1947.

Paradis, J.G., and M.L. Zimmerman, *The MIT Guide to Science and Engineering Communication*, Cambridge, MA: MIT Press, 1997.

Patton, W.L., *An Author's Guide to the Copyright Law*, Lexington, MA: Lexington Books, 1980.

Pauley, S.E., and D.G. Riordan, *Technical Report Writing Today*, Boston: Houghton Mifflin Co., 1990.

Perelman, L.C., J. Paradis, and E. Barrett, *The Mayfield Handbook of Technical & Scientific Writing*, Mountain View, CA: Mayfield Publishing Co., 1998.

Peterson, M.S., *Scientific Thinking and Scientific Writing*, New York: Reinhold Publishing Corp., 1961.

Raman, M., and S. Sharma, *Technical Communication: Principles and Practice*, New York: Oxford University Press, 2004.

Rathbone, R.R., *Communicating Technical Information: A New Guide to Current Uses and Abuses in Scientific and Engineering Writing*, Reading, MA: Addison-Wesley, 1985.

Reisman, S.J., *A Style Manual for Technical Writers and Editors*, New York: Macmillan, 1962.

Rhodes, F.H., *Technical Report Writing*, New York: McGraw-Hill Book Company, Inc., 1941.

Riney, L.A., *Technical Writing for Industry: An Operations Manual for the Technical Writer*, Englewood Cliffs, NJ: Prentice Hall, 1989.

Riordan, D.G., *Technical Report Writing Today*, Boston: Houghton Mifflin, 2002.

Robertson, W.S., and W.D. Siddle, *Technical Writing & Presentation*, Oxford, NY: Pergamon Press, 1966.

Rubens, P., General Ed., *Science and Technical Writing: A Manual of Style*, New York: H. Holt, 1992.

Sandman, P.M., C.S. Klompus, and B.G. Yarrison, *Scientific and Technical Writing*, New York: Holt, Rinehart, and Winston, 1985.

Schmidt, S., *Creating the Technical Report*, Englewood Cliffs, NJ: Prentice-Hall, 1983.

Schoenfeld, R., *The Chemist's English*, Deerfield Beach, FL: VCH Publishers, 1985.

Sherman, T.A., *Modern Technical Writing*, Englewood Cliffs, NJ: Prentice-Hall, 1957.

Sides, C.H., *How To Write & Present Technical Information*, Phoenix: Oryx Press, 1999.

Souther, J.W., and M.L. White, *Technical Report Writing*, New York: John Wiley & Sons, 1977.

Sprent, P., *Getting into Print: A Guide for Scientists and Technologists*, New York: E. & F.N. Spon, 1995.

Trzyna, T.N., and M.W. Batschelet, *Writing for the Technical Professions*, Belmont, CA: Wadsworth Publishing Co., 1987.

Ulman, J.N., *Technical Reporting*, New York: Holt, 1952.

van Emden, J., and J. Easteal, *Technical Writing and Speaking*, New York: McGraw-Hill, 1996.

Van Hagan, C.E., *Report Writers' Handbook*, New York: Dover Publications, Inc., 1961.

Wallace, J.D., and J.B. Holding, *Guide to Writing and Style*, Columbus, OH: Battelle Memorial Institute, Columbus Laboratories, 1966.

Ward, R.R., and R.M. Ohman, Ed., *Practical Technical Writing*, New York: Knopf, 1968.

Weil, B.H., *The Technical Report: Its Preparation, Processing, and Use in Industry and Government*, New York: Reinhold Publishing Corp., 1954.

Weisman, H.M., *Technical Report Writing*, Columbus, OH: Merrill, 1975.

Weiss, E.H., *101 Writing Remedies: Practical Exercises for Technical Writing*, Phoenix: Oryx Press, 1990.

Winsor, D.A., *Writing Like an Engineer: A Rhetorical Education*, Mahwah, NJ: Lawrence Erlbaum Associates, 1996.

Wolcott, H.F., *Writing Up Qualitative Research*, Newbury Park, CA: Sage Publications, 1990.

Woolston, D.C., P.A. Robinson, and G. Kutzbach, *Effective Writing Strategies for Engineers and Scientists*, Chelsea, MI: Lewis Publishers, 1988.

Zall, P.M., *Elements of Technical Report Writing*, New York: Harper, 1962.

Zimmerman, D.E., and D.G. Clark, *The Random House Guide to Technical and Scientific Communication*, New York: Random House, 1987.

# *Abbreviations*

| | |
|---|---|
| AD | Accessioned Document |
| AFRL | Air Force Research Laboratory |
| ANSI | American National Standards Institute |
| APL | The Johns Hopkins University Applied Physics Laboratory |
| C | Confidential |
| CD-ROM | Compact Disk–Read-Only Memory |
| CNWDI | Critical Nuclear Weapon Design Information |
| COMSEC | Communications Security |
| COTR | Contracting Officer's Technical Representative |
| CR | Contractor Report |
| DDLRS | Defense Digital Library Research Service |
| DEA | Drug Enforcement Administration |
| DFARS | Defense Federal Acquisition Regulation Supplement |
| DID | Data Item Description |

| | |
|---|---|
| DOD/DoD | Department of Defense |
| DoD UCNI | DoD Unclassified Controlled Nuclear Information |
| DODD | Department of Defense Directive |
| DoN | Department of the Navy |
| DSP | Defense Standardization Program |
| DSS | Defense Security Service |
| DTIC | Defense Technical Information Center |
| EO | Executive Order |
| FAR | Federal Acquisition Regulation |
| FFRDC | Federally Funded Research and Development Center |
| FOIA | Freedom of Information Act |
| FOUO | For Official Use Only |
| FRD | Formerly Restricted Data |
| GIDEP | Government-Industry Data Exchange Program |
| GPR | Government Purpose Rights |
| IAC | Information Analysis Center |
| IR&D | Independent Research and Development |
| ISCAP | Interagency Security Classification Appeals Panel |
| ISO | International Organization for Standardization |
| ISOO | Information Security Oversight Office |
| JCP | Joint Certification Program |
| LH | Lower Half (Rear Admiral) |
| MCTL | Militarily Critical Technologies List |
| N | Critical Nuclear Weapon Design Information |
| NATO | North Atlantic Treaty Organization |
| NAVSEA | Naval Sea Systems Command |
| NAVSEASYSCOM | Naval Sea Systems Command |
| NC | Not Releasable to Contractors/Consultants |
| NF | Not Releasable to Foreign Nationals |
| NIPRNet | Non-secure Internet Protocol Router Network |

| | |
|---|---|
| NISO | National Information Standards Organization |
| NNPI | Naval Nuclear Propulsion Information |
| NOCONTRACT | Not Releasable to Contractors/Consultants |
| NOFORN | Not Releasable to Foreign Nationals |
| NSWCCD | Naval Surface Warfare Center, Carderock Division |
| NTIS | National Technical Information Service |
| OADR | Originating Agency's Determination Required |
| OC | Dissemination and Extraction of Information Controlled by Originator |
| OPNAVINST | Office of the Chief of Naval Operations Instruction |
| ORCON | Dissemination and Extraction of Information Controlled by Originator |
| PDF | Portable Document Format |
| PMO | Program Management Office |
| PR | Caution—Proprietary Information Involved |
| PROPIN | Caution—Proprietary Information Involved |
| R&D | Research and Development |
| RD | Restricted Data |
| RDDS | Research and Development Descriptive Summaries |
| RDT&E | Research, Development, Test, and Evaluation |
| R&E | Research and Engineering |
| REL TO | Authorized for Release to |
| RS | Research Summaries |
| S | Secret |
| SAR | Same as Report |
| SBIR | Small Business Innovation Research |
| SBU | Sensitive But Unclassified |
| SECNAVINST | Secretary of the Navy Instruction |
| SIPRNet | Secret Internet Protocol Router Network |
| SNM | Special Nuclear Material |
| STI | Scientific and Technical Information |

| | |
|---|---|
| STINET | Scientific and Technical Information Network |
| STIP | Scientific and Technical Information Program |
| STRN | Standard Technical Report Number |
| TAT | Technical Area Task |
| TIC | Technical Information Center |
| TM | Technical Memorandum |
| TR | Technical Report |
| TS | Top Secret |
| U | Unclassified |
| UARC | University-Affiliated Research Center |
| UH | Upper Half (Rear Admiral) |
| URL | Uniform Resource Locator |
| U.S.C. | U.S. Code |
| USSAN | United States Security Authority, NATO |
| UU | Unclassified Unlimited |

# *Glossary*

Appendix B of ANSI/NISO Z39.18-2005[ac] is a glossary. Terms defined in Appendix B of ANSI/NISO Z39.18-2005 are not repeated in this glossary. Unless indicated otherwise, the terms and definitions of terms in this glossary are original, common, or were obtained from various Department of Defense (DoD) sources. Some DoD terms and definitions have been modified to apply to the subject matter described herein.

**Academia.** A university-affiliated research center (UARC), such as the Johns Hopkins University Applied Physics Laboratory, and a federally funded research and development center (FFRDC), such as the Lincoln Laboratory at the Massachusetts Institute of Technology.

**All-capital letters.** All letters in a word are uppercase. (See also **Initial capital letters**.)

**Ancillary information (figure and table titles).** A supplementary, clarifying parenthetical phrase or sentence accompanying a figure or table title. Ancillary information should not be included in the list(s) of figures and tables.

---

[ac]National Information Standards Organization, "Scientific and Technical Reports - Preparation, Presentation, and Preservation," ANSI/NISO Z39.18-2005, Bethesda, MD: NISO, 2005.

**Boldface type.** **This is an example of boldface type.** (See also **Light-face type.**)

**Callout.** An identifying symbol, word, or group of symbols and/or words in a figure. A callout is not the figure title or subtitle.

**Cell (table).** Where columns and rows intersect and where data are inserted. (See also **Column [table]; Row [table].**)

**Classification by compilation.** Individually unclassified items of information that may become classified if the compiled information reveals an additional association or relationship. The compilation must qualify for classification under an Executive order.

**Column (table).** Vertical component that intersects rows and contains cells. (See also **Cell [table]; Row [table].**)

**Contracting officer.** A military officer or civilian employee who has a valid appointment as a contracting officer under the provisions of the Federal Acquisition Regulation. The individual has the authority to enter into and administer contracts and determinations, as well as findings about such contracts.

**Contracting Officer's Technical Representative (COTR).** An individual appointed by the contracting officer who represents the contracting officer on technical matters.

**Contractor.** An individual/organization outside the Government who/ which has agreed to provide research, supplies, or services to a Government agency. This includes prime contractors and subcontractors from academia and industry.

**Controlling DoD office.** The sponsoring DoD activity that generated the technical report or received the technical report on behalf of DoD and, therefore, has responsibility for determining the distribution availability of the technical report. For joint sponsorship, the controlling DoD office is determined by advance agreement and may be a party, group, or committee representing the interested activities or DoD components.

**Cover sheet.** The first page of some appendixes containing the appendix number and title. The back of the cover sheet is blank.

**Critical nuclear weapon design information (CNWDI).** A category of weapon data designating Top Secret Restricted Data or Secret Restricted Data revealing the theory of operation or design of the components of a thermonuclear or implosion-type fission bomb, warhead, demolition munition, or test device. Access to CNWDI is on a need-to-know basis.

**Data item description (DID).** A DID is a standardization document that defines the data content, preparation instructions, format, and intended use of data required of a contractor. DIDs should be prepared in accordance with MIL-STD-963.[ad]

**Decimal numbering subordination and format.** One of two types of formats specified for technical reports, whereby decimal numbering, i.e., headings and subheadings are assigned a decimal number, is used in lieu of typographical progression to indicate subordination of headings. This type of subordination is more conducive to large, complex technical reports. (See also **Typographical progression subordination and format.**)

**Defense.** The defense of the United States and its Allies. However, "defense" is not confined to strictly military-related topics, such as weaponry; it also includes subjects that, oftentimes, have military and civilian applications, referred to as "dual-use technologies."

**Defense Technical Information Center (DTIC).** DoD's central repository for research, development, test, and evaluation (RDT&E) information in all fields of defense-related science and technology.

**Department of Defense components.** The Department of the Army; the Department of the Navy, including the Marine Corps; and the Department of the Air Force.

**Desktop publishing.** Electronically designing, laying out, editing, and producing a document using a personal computer and appropriate software.

**Distribution statement.** A required statement used to mark a technical report to denote the extent of its availability for distribution, release, and disclosure.

**Em dash/space.** A dash or space as wide as the point size of a particular font (roughly equal to the width of a capital "M"); twice the width of an en dash/space.

**En dash/space.** A dash or space half as wide as the point size of a particular font (roughly equal to the width of a capital "N"); one-half the width of an em dash/space.

**Equation, displayed.** An equation appearing on a separate line, as opposed to appearing within the text. Displayed equations are usually mentioned in the text and numbered.

**Flush left/right.** Aligned with the left/right margin of a page.

---

[ad]MIL-STD-963, "Department of Defense Standard Practice: Data Item Descriptions (DIDs)," 31 August 1997.

**Font.** A complete set of type of one size and face. (See also **Typeface.**)

**Footer.** A continuous design element and/or set of information at the bottom of each page or table. (See also **Header.**)

**Government.** The U.S. Government, unless stated otherwise.

**Government-Industry Data Exchange Program (GIDEP).** GIDEP was established to eliminate duplicate work in engineering, reliability and maintainability, failure experience, and metrology.

**Hanging indentation.** A type of indentation where the first line is flush left and runover lines are uniformly indented. The glossary entries herein are examples of a hanging indentation.

**Header.** A design element and/or set of information at the top of each page or table. (See also **Footer.**)

**Helvetica.** A sans serif typeface. This is an example of Helvetica typeface.

**Industry.** DoD contractors (and subcontractors), such as Lockheed Martin and Northrop Grumman.

**Inferior (equation).** Subscript.

**Initial capital letters.** The first letter in a word is uppercase; the remaining letters are lowercase. In titles and headings, the first word and each subsequent word begin with an uppercase letter, except articles and prepositions of four or less letters within the series of words. Infinitives also begin with an uppercase letter. For example, "The Technical Report To Be Submitted for Review."

**Initial indentation.** The indentation of the first line of a multiple-line paragraph.

**Inverted pyramid style arrangement.** Arrangement of headings and figure or table titles resembling an inverted pyramid, where each line is slightly longer than the line below it. (See also **Pyramid style arrangement.**)

**Italic type.** Slanted type. *This is an example of italic type.* (See also **Roman type.**)

**Justified.** Aligned evenly with the left and right margins or the top and bottom margins. (See also **Ragged edges.**)

**Landscape orientation.** Horizontally oriented on a page. (See also **Portrait orientation.**)

**Layout.** The arrangement of text, figures and tables, and displayed equations on a page.

**Leaders.** Dots in a row leading the eye across a page. A minimum of three leaders should be inserted in a table of contents and list(s) of figures and tables between an entry and its corresponding page number.

**Legend.** An explanation usually accompanying a figure or table explaining the use of certain symbols or text.

**Lightface type.** This is an example of lightface type. (See also **Boldface type.**)

**Line length.** The horizontal measure of a line of text.

**Lowercase letter.** A letter that is not a capital or uppercase letter. (See also **Uppercase letter.**)

**Margins.** The areas around the edges of a page, which form a boundary for text and figures and tables, as well as additional areas applied to text and figures and tables. A page has a top margin, a bottom margin, and inner (binding edge) and outer margins. Text and figures and tables have a top margin, a bottom margin, and left and right margins.

**Multiple sources.** Two or more source documents, classification guides, or a combination of both.

**Parallel structure.** A principle whereby similar words or ideas are expressed in a similar manner, e.g., heat*ing*, ventilat*ing*, and air condition*ing*; *not* heat*ing*, ventilat*ion*, and air condition*ing*. Parallel structure is most easily recognized in lists.

**Pica.** Approximately 1/6 in.

**Point.** Approximately 1/72 in.; 12 points equals 1 pica.

**Portion.** Heading, paragraph, and, possibly, list and equation in a classified technical report. Portions require classification markings. Portions are also marked in unclassified For Official Use Only (FOUO) technical reports.

**Portrait orientation.** Vertically oriented on a page. If possible, text and figures and tables should appear in a portrait orientation. (See also **Landscape orientation.**)

**Primary distribution.** Initial targeted distribution of or access to technical reports authorized by the controlling DoD office. (See also **Secondary distribution.**)

**Proprietary information.** Intellectual property that a private firm has developed at its own expense and that it wants to control.

**Public release.** Making a technical report available to the public without restricting its dissemination or use. This includes foreign disclosure.

**Pyramid style arrangement.** Arrangement of headings and figure or table titles resembling a pyramid, where each line is slightly shorter than the line below it. (See also **Inverted pyramid style arrangement.**)

**Ragged edges.** Paragraphs having an uneven line length with the right margin. Ragged edges facilitate reading. (See also **Justified.**)

**Research.** All efforts directed toward increased knowledge of natural phenomena and the environment and efforts directed toward the solution of long-term defense problems in the physical, engineering, life, behavioral, and social sciences.

**Roman type.** Upright type. This is an example of roman type. (See also **Italic type.**)

**Row (table).** Horizontal component that intersects columns and contains cells. See also **Cell; Column (table).**

**Rule.** A line drawn on a page, often associated with a footnote.

**Runover.** Text continuing to the next line.

**Sans serif.** Typeface without serifs or tails, e.g., Helvetica typeface. (See also **Serif.**)

**Scientific and technical information (STI).** Communicable knowledge or information resulting from or pertaining to the conduct and management of RDT&E efforts. STI advances the state of the art or establishes a new art in area of significant military application. Included is information related to munitions and other military supplies and equipment.

**Scientific and Technical Information Program (DoD's).** A coordinated structure of STI functions. The objective of the program is to ensure that STI makes the best possible contribution toward the advancement of science and technology by permitting the timely, effective, and efficient conduct of DoD RDT&E programs; providing information to support management of RDT&E-related programs; and eliminating unnecessary duplication of effort and resources by encouraging and expediting the interchange and use of STI.

**Scientific and technical reports.** Scientific and technical reports traditionally "convey the results of basic or applied research and support decisions based on those results. A [scientific and technical] report includes the ancillary information necessary for interpreting, applying, and replicating the results or techniques of an investigation. The primary purposes of such a report are to disseminate the results of scientific and technical research and to recommend action."[ae] The formats prescribed herein adhere to this definition and logic. Scientific and technical reports may be "print"[af] (paper and ink) oriented or they may be "non-print"[af] (totally digital) in nature.

**Secondary distribution.** Release of a technical report after primary distribution. This includes allowing it to be read only, loaning it, or releasing it in whole or in part. (See also **Primary distribution**.)

**Separate document.** A document created prior to and exclusive of a technical report. Unpublished separate documents or portions thereof may be appended to technical reports.

**Serif.** Typeface with distinctive tails on vertical and horizontal lines, e.g., Times typeface. (See also **Sans serif**.)

**Spine.** Bound edge of a document.

**Standalone element.** An element within a technical report, e.g. executive summary, that is meant to be read and understood in its entirety without reference to or reliance upon other elements of the technical report or to other documents.

**Standardization.** The process by which DoD achieves the closest practicable cooperation among the services and defense agencies for the most efficient use of research, development, and production resources and agrees to adopt on the broadest possible basis the use of common or compatible operational, administrative, and logistic procedures; common or compatible technical procedures and criteria; common, compatible, or interchangeable supplies, components, weapons, or equipment; and common or compatible tactical doctrine with corresponding organizational compatibility.

**Style.** A particular manner of written expression involving choices or options that include capitalization, spelling, word compounding, punctuation, abbreviations and letter symbols, italic, and numerals. (See also **Tone**.)

**Subscript.** Character written next to and slightly below a letter or number. (See also **Superscript**.)

**Superior (equation).** Superscript.

---

[ae]National Information Standards Organization, "Scientific and Technical Reports—Elements, Organization, and Design," ANSI/NISO Z39.18-1995, Bethesda, MD: NISO, 1995.

[af]National Information Standards Organization, "Scientific and Technical Reports - Preparation, Presentation, and Preservation," ANSI/NISO Z39.18-2005, Bethesda, MD: NISO, 2005.

**Superscript.** A character written next to and slightly above a letter or number. (see also **Subscript.**)

**Technical information center (TIC).** Any activity that acquires, organizes, houses, retrieves, and disseminates information and information materials and performs reference and research in direct support of an organization's RDT&E mission. A TIC may also provide all or any one of such services as analysis, current awareness, literature search, translations, and referral.

**Technology.** Scientific or engineering efforts directed toward eliminating technical barriers and providing solutions to technical problems (excluding routine engineering) encountered in RDT&E programs.

**Times.** A serif typeface. This is an example of Times typeface.

**Tone.** A particular manner of written expression involving the attitude or feeling conveyed by the writer (author) toward the reader. There are basically two types of tone: formal and informal. Defense-related technical reports should adopt a formal tone. (See also **Style.**)

**Turnpage.** A page oriented vertically (portrait) with regard to the page header and footer and containing a figure or table oriented horizontally (landscape). See also **Landscape orientation** and **Portrait orientation**.

**Typeface.** A full range of type of the same design e.g., Times is a specific typeface of the serif typefaces; Helvetica is a specific typeface of the sans serif typefaces. (See also **Font.**)

**Typographical progression subordination and format.** One of two types of formats specified for technical reports, whereby typographical progression, i.e., the location of headings on a page and the use of boldface and italic type or combinations thereof in headings, is used in lieu of decimal numbering to indicate subordination of headings. This type of format is more conducive to short, less complex technical reports. (See also **Decimal numbering subordination and format.**)

**Typography.** The arrangement and appearance of printed matter.

**Uppercase letter.** A capital letter. See also **Lowercase letter**.

**Visuals.** Figures and tables. Visuals do not include equations.

# *Index*

## A

abbreviations
  chemical, **298**, 298
  classified technical reports, 26
  description, 265
  list of, 205
  standard, **267**, 267, **286**
  title and subtitle, 73
abstract
  ANSI/NISO Z39.14-1997, 7
  audience, 245
  classified technical reports, 33
  color paper, 56
  comparison with executive
    summary, 146, 147
  description, 117
  examples, *120*, *121*

abstract (continued)
  order of appearance, 65
  report documentation page
    (Standard Form 298), 116
  standalone element, 123
  visuals, 34
academic degrees, 75
acknowledgments
  description, 140
  examples, *142*, *143*
  order of appearance, 65
  relationship to preface, 140
acronyms
  abstract, 118
  description, 265
  keywords, 119
  list of, 205
  list(s) of figures and tables, 132
  table of contents, 123

acronyms (continued)
  title and subtitle, 73
AD–A 423 966, 85
administrative information. *See*
  preface
administrative/operational use, 88
aircraft, 300
American National Standards
  Institute (ANSI). *See specific
  standard*
ANSI Z39.18-1987, 4
ANSI/NISO Z39.14-1997, 7, 117
ANSI/NISO Z39.18-1995, 5, 24,
  248
ANSI/NISO Z39.18-2005
  abstracts, 117
  acknowledgments, 140
  appendixes, 187
  authors, 74
  back cover, 68
  back matter of technical report,
    187
  bibliography, 192
  body of technical report, 145
  conclusions, 158
  copyrights, 68
  description, 5
  disclaimers, 67
  distribution list, 219
  equations, 57
  executive summary, 146
  figures, 36, 44, 47
  footnotes, 36
  foreword, 133
  front matter, 65
  glossary, 209
  image area, 52
  index, 213
  introduction, 149
  list(s) of figures and tables, 125

ANSI/NISO Z39.18-2005
  (continued)
  methods, assumptions, and
    procedures, 156
  organization of technical
    reports, 24
  page column format and line
    length, 50
  preface, 138
  publication of, xxiv
  recommendations, 158
  references, list of, 167
  report documentation page
    (Standard Form 298), 117
  results and discussion, 157
  subordination in technical
    reports, 25
  symbols, abbreviations, and
    acronyms, list of, 205
  table of contents, 119
  tables, 36
  typography, 52
ANSI/NISO Z39.23-1997, 7, 69
Anti-Deficiency Act, 235
appendixes
  description, 187
  examples
    classified technical report
      decimal numbering format,
        *199, 200, 202*
      typographical progression
        format, *192, 193, 195*
    unclassified technical report,
      *190, 197*
  list(s) of figures and tables, 132
  order of appearance, 187
  table of contents, 123
  use of, 245
  visuals, 34, 36
applied research, 13

Arms Export Control Act, 92
atomic energy information, 83, 94, 96
audience for technical reports, 245
author(s)
    appearance in foreword, 133
    audience for book, xxiii
    classified technical reports, 21
    description, 73
    distribution list, 85
    front cover and title page, 66
    publishing requirements, 9
    relationship to performing organization, 245
    report documentation page (Standard Form 298), 108

**B**

back matter of technical report, 187
basic research, 12, 89
bibliography
    comparison with list of references, 167
    description, 192
    examples, *204-207*
    order of appearance, 187
blank pages, 57
body of technical report
    description, 145
    examples, *159, 163, 167, 171*
boldface type, 246
brand names, 67

**C**

capitalization, 247, 249, 251
caution (technical manual), 2, *3*
chemical names and abbreviations, **298**, 298

*Chicago Manual of Style, The*, 304
civilian job titles, 75
classified technical reports
    abstract, 33, 117, 119, *121*
    acknowledgments, 141, *143*
    appendixes
        examples
            body of unclassified appendix, *193, 200*
            classified appendix, *195, 202*
            cover sheet of unclassified appendix, *192, 199*
        format, 189
    author responsibilities, 21
    availability on STINET, 234, 244
    back cover, 68, *111*
    back of title page, *109*
    bibliography, 199, *205, 207*
    blank pages, 57
    body of technical report, *163, 171*
    compilation of information, 56
    DEA sensitive information, 80
    description, 20
    destruction, 98
    disclaimers, 67
    distribution
        contractor restrictions, 16
        distribution statement limitations, 89, 90
        foreign national restrictions, 20
        GIDEP prohibition, 16
        preparing organization responsibility, 15
    distribution list, 223, *225, 227*
    distribution statements, 85, 87
    document control number, 71

classified technical reports
(continued)
downgrade, declassify, and
related markings, 93
equations, 59, *60*
errata, 62, *63*
examples. *See specific example*
executive summary, 33, 147,
*150, 154*
export control notice, 91
figures, *48*
footnotes, 38, 61, *62*
foreign government and NATO
information, 99
foreword, 136, *143*
FOUO information, 79
front cover, *105*
glossary, 209, *215, 217*
headings
appearance, 27, 28, 33
examples, *29, 33*
index, 218, *220, 222*
inside front cover, 68, *107*
intelligence control markings,
81
list(s) of figures and tables
description, 132
examples, *135, 136, 138, 139*
lists, 30, *32*, 34, *35*
mailing, 17
markings
basic requirements, 22, 68
description, 25
noncommercial data and
software, 84
report documentation page
(Standard Form 298), 116
original and derivative
classification, 22
overall classification, 66, 76
page format, 50, 53, *55*, 56

classified technical reports
(continued)
page numbers, 53
paragraphs, 28
preface, 140, *143*
reasons for classifying, 21
references, list of, 176, *178,
180*, **181**
report documentation page
(Standard Form 298), 102,
107
security review, 11
submission to DTIC, 239
symbols, abbreviations, and
acronyms, list of, 208, *211,
213*
table of contents
description, 123
examples, *127, 128, 130, 131*
tables, *51*
title and subtitle, 73
title page, *106*
unclassified controlled
information, 77
visuals
examples, *40, 43, 45*
markings, 38
page classification, 41
warning notices, 83
Commerce, Department of, 14, 91,
235
Commerce Control List, 14, 91
Communications Security
(COMSEC) material, 83
compounding rules, 264
Computer Security Act of 1987,
77, 81
conclusions, 145, 157
consultants. *See* contractors and
consultants
contract number, 107, 139

Figures - *Italic*; Tables - **Boldface**

Contract Security Classification
  Specification, 93, 94, 96
contracting officer's technical
  representative (COTR), 10, 89
contractions, 247
contractors and consultants
  COMSEC material, 83
  intelligence control markings,
    82, 115
  performance evaluation, 88
  technical reports
    authorship, 74
    back of title page, 69
    cover and title page, 102
    inside front cover, 67
    seals/emblems and logos, 101
controlling office, 89
copyrights
  authors, 74
  description, 68
  figures and tables, 176
  notices, 67, 84
  report documentation page
    (Standard Form 298), 109
corporate author, 74, 245, 246
cover
  *See also* title page
  back
    DEA sensitive information,
      80
    description, 68
    DoD UCNI, 81
    examples, *110, 111*
    mention, 223
    order of appearance, 65, 187
    overall classification, 76
    seals/emblems and logos, 101
    technical report number, 70
    unclassified controlled
      information, 77

cover (continued)
  back (continued)
    unclassified sensitive
      information, 81
  front
    authors, 74, 75
    DEA sensitive information,
      80
    description, 65
    distribution statement, 85
    DoD UCNI, 81
    downgrade, declassify, and
      related markings, 93
    examples, *103, 105*
    export control notice, 91
    foreign government and
      NATO information, 99
    mention, 108, 223
    notice, 56, 57
    order of appearance, 65
    overall classification, 76
    restrictive markings on
      noncommercial data and
      software, 66, 84
    seals/emblems and logos, 101
    technical report number, 70
    unclassified controlled
      information, 77
    unclassified sensitive
      information, 81
    warning notices, 82
  inside back, 68
  inside front, 65, 67, *107*
cover sheet on appendixes, 189,
  *192, 199*
Critical Nuclear Weapon Design
  Information (CNWDI), 83
critical technology, 87, 91

# D

data item description (DID), 10
data rights, Government, 84
date of determination, 89
decimal numbering format
    *See also* page format;
        typographical progression
        format
    abstract, *120, 121*
    ANSI/NISO Z39.18-2005
        specification, xxvi
    appendixes
        description, 188
        examples, *197, 199, 200, 202*
    bibliography, *206, 207*
    body of technical report, *167,
        171*
    description, 31
    distribution list, *226, 227*
    errata, 62, *63*
    examples. *See specific example*
    executive summary, 147, *152,
        154*
    footnotes, 61
    glossary, 210, *216, 217*
    headings, 31, *33*
    index, *221, 222*
    introduction, 156
    list(s) of figures and tables, 133,
        *137-139*
    lists, 34, *35*
    page format, 53
    pages, 50
    paragraphs, 34
    references, list of, 166, *179, 180*
    symbols, abbreviations, and
        acronyms, list of, 209, *212,
        213*
    table of contents, 125, *129-131*
    visuals, 37

Defense (DoD), Department of
    *See also* Defense Technical
        Information Center (DTIC);
        *other specific agencies and
        subject matter*
    audience for book, xxiii
    authors, 73
    Director, Defense Research and
        Engineering, 233
    Militarily Critical Technologies
        List, 11, 14
    scientific and technical
        information program, 8, 236
    standards for technical reports, 4
    Unclassified Controlled Nuclear
        Information, 77, 81
    Under Secretary of Defense for
        Acquisition, Technology and
        Logistics, 233
Defense Digital Library Research
    Service (DDLRS), 244
Defense Federal Acquisition
    Regulation Supplement
    (DFARS), 69, 84, 235
Defense Logistics Agency, 16
Defense Security Service (DSS),
    16
Defense Technical Information
    Center (DTIC)
    *See also* libraries
    abstracts, 118
    accessioned document (AD)
        number, 175, 238
    central repository for STI, xxv
    core collections, *237*
    distribution list, 221
    distribution of technical reports,
        16, 85, 235, 242
    Form 55, 17, *18, 19*, 236
    Handle Service, 244

Defense Technical Information
    Center (DTIC) (continued)
  independent research and
    development (IR&D)
    database, 239
  Information Analysis Centers
    (IACs), 234, 240
  literature search, 156
  mailing address, 239
  Militarily Critical Technologies
    List (MCTL), 92
  overview, 233
  QuestionPoint ("Ask a
    Librarian"), 244
  registrants, 242
  registration with, 242
  report documentation page
    (Standard Form 298), 102
  Research and Development
    Descriptive Summaries
    (RDDS) database, 240
  research at, 243
  research summaries (RS)
    database, 239
  Scientific and Technical
    Information Network
    (STINET), 234, 243
  submission guidelines, 238
  technical report (TR) database,
    4, 175, 236, 238
  technical review process, 11
definition of a technical report, 1
Department. *See other part of title*
design of technical reports, 25
destruction notice, 66, 98
Developing Science and
    Technologies List, 92
DFARS 227.71, 84
DFARS 227.72, 84
DFARS 235.010, 235
DFARS 252.227-7013, 84

DFARS 252.227-7014, 84
DFARS 252.227-7018, 84
DFARS 252.235-7011, 235
digital-access information, 169
digital technical reports, 6
disclaimers, 67
distribution and duplication notice,
    83
distribution list
  description, 218
  document control number, 71
  examples, *224-227*
  mention, 85
  order of appearance, 187
distribution of technical reports
  basic guidelines, 15
  external distribution, 16
  foreign distribution, 17
  internal distribution, 16
  primary distribution, 15, 61, 85
  secondary distribution, 17, 85,
    235
distribution statements
  checklist, **86**
  date of determination, 89
  definitions of reasons for
    limiting distribution of
    technical reports, 87
  description, 85, 89
  examples, **85**
  front cover and title page, 66
  noncommercial data and
    software, 84
  report documentation page
    (Standard Form 298), 109
  review, 12
  selection, 10
  use of, 157
document control number, 66, 71
DoD 3200.12, 235, 236, 238
DOD 5200.1-PH, 76, 93, 95

Figures - *Italic*; Tables - **Boldface**

DOD 5200.1-R
  destruction notice, 98
  distribution statement F, 90
  foreign government and NATO
    information, 99
  reference, 76
  unclassified controlled
    information, 77
DoD 5210.2, 83
DOD 5220.22-M, 93, 98
DoD 5230.25, 88, 90, 92
DoD Unclassified Controlled
  Nuclear Information (DoD
  UCNI), 77, 81
DODD 5230.9, xxv, 89
DODD 5230.11, 82
DODD 5230.24, 85
downgrade, declassify, and related
  markings, 93, **96**
Drug Enforcement Administration
  (DEA), 77, 80

**E**

editor's and illustrator's marks,
  249, *250*
editors
  acknowledgments, 140
  audience for book, xxiii
  mention, 23, 74
  preface, 137, 140
  responsibilities, 9
  review of technical reports, 11
elements of a technical report, **24**
emblems/seals, 66, 101
emphatic phrases, 246
Energy (DoE), Department of,
  xxiv

equations, 26, 34, 57, *60*
errata, 61, *63*
exclamatory sentences, 247
Executive Order 12958, 95, **96**
Executive Order 13292
  description of classified
    information, 20
  downgrade, declassify, and
    related markings, 93, 94, **96**
executive summary
  audience, 245
  classified technical reports, 33
  color paper, 56
  comparison with abstract, 117
  description, 146
  examples, *148, 150, 152, 154*
  length, 245
  order of appearance, 145
  recommendations, 158
  standalone element, 123, 147
  table of contents, 123, 132
  visuals, 34, 36
Export Administration Act of
  1979, 92
Export Administration
  Regulations, 92
export-controlled information
  determination, 10
  distribution
    contractor eligibility, 16
    critical technology, 88
    direct military support, 88
    distribution statement A, 89
    distribution statement X, 90
    distribution statements B, C,
      D, and E, 90
    DTIC Form 55, 17
    NNPI, 83
    notice, 66, 82, 91

Figures - *Italic*; Tables - **Boldface**

export-controlled information
(continued)
review, 14

# F

figures
*See also* tables; visuals
appendixes, 189
callouts, 47
classification markings, 26, *48*
color, 47
complementary role, 247
components, *48*
description, 41
errata, 62
examples. *See specific example*
footnotes, 47, 61
indexing, 218
information absorption, *46*
information retention, *46*
references, 176
results and discussion, 157
standalone elements, 132
titles, 26
footnotes
citations, 162
classification markings, 26
classified technical reports, 38
description, 36, 61
examples, *62*
figures, 47
tables, 49
For Official Use Only (FOUO)
information, 77, 98
foreign nationals, organizations,
and governments
abbreviations, **114**, 114
classification levels, **99**

foreign nationals, organizations,
and governments (continued)
distribution of technical reports,
89
information, 87, 99
intelligence control markings,
82, 115
NNPI, 83
foreword
comparison to
acknowledgments, 140
description, 133
examples, *142, 143*
order of appearance, 65
format and subordination of
technical reports
*See also* decimal numbering
format; page format;
typographical progression
format
description, 25
Freedom of Information Act
(FOIA), 77, 235, 236
front matter of technical report, 65

# G

gender-oriented language, 246
Global Reference Network, 244
glossary
description, 209
examples, *214-217*
order of appearance, 187
Government data rights, 84
Government-Industry Data
Exchange Program (GIDEP), 16
Government Purpose Rights, 84
grant number, 107, 139

# H

Handle Service, 244

headings
    classification markings, 26
    description, 27, 31
    examples, *29, 33*
    typography, 52
Homeland Security (DHS),
    Department of, xxiv

## I

illustrators
    *See also* editors
    acknowledgments, 140
    audience for book, xxiii
    mention, 23, 74
    responsibilities, 9
independent research and
        development (IR&D)
    DTIC database, 239
    mention, 108
    review of technical reports, 11
    STINET, 243, 244
index
    description, 213
    examples, *219-222*
    order of appearance, 187
informal terms, 248
Information Analysis Centers
    (IACs), 234, 240
initialisms, 265
intelligence control markings
    abbreviations, 26, 27
    description, 81
    front cover and title page, 66
    report documentation page
        (Standard Form 298), 109
intelligence sources and methods,
    82
Interagency Security Classification
    Appeals Panel (ISCAP), 95

International Exchange Program,
    88
International Organization for
    Standardization (ISO), 15
International Traffic in Arms
    Regulations, 92
introduction, 118, 145, 147
italic, 246, 300

## J

jargon, 246
Joint Certification Program (JCP),
    U.S./Canada, 17
Justice, Department of, 77

## K

keywords, 116, 118

## L

leaders (format), 123, 132
legal counsels, 14, 16
letter designations, 301
librarians, xxiii
libraries
    *See also* Defense Technical
        Information Center (DTIC)
    abstracts, 118
    distribution list, 222
    internal distribution, 16
    literature search, 156
    QuestionPoint ("Ask a
        Librarian"), 244
    report documentation page
        (Standard Form 298), 102
    technical review process, 11
list(s) of figures and tables
    *See also* table of contents
    consistency with table of
        contents, 125

list(s) of figures and tables
    (continued)
  description, 125
  examples
    classified technical report,
      *135*, *136*, *138*, *139*
    unclassified technical report,
      *134*, *137*
  order of appearance, 65
lists
  classification markings, 26
  description, 30, 34
  examples, *32*, *35*
  recommendations, 158
logos, 66, 101

## M

measure, terms of, 285
measurement, units of, **286**, 286
measurements, 156
memorandums, 3, 61
metadata, 5
methods, assumptions, and
    procedures, 145, 156
MIL–STD–847, 4
MIL–STD–38784, 2, 32
Militarily Critical Technologies
    List, 11, 14, 92
military or space applications, 91
military organizations, names of,
    249
military titles, **75**, 75, 264
multimedia, 7, 23
Munitions List, U.S., 14, 91

## N

National Information Standards
    Organization (NISO)
  *See also specific standard*
  metadata description, 5
  standards development, 4
National Institute of Standards and
    Technology (NIST), 114
National Technical Information
    Service (NTIS), 175, 235
Naval Nuclear Propulsion
    Information (NNPI), 83
Navy (DoN), Department of the,
    xxvi
noncommercial data, 66, 84
non-print technical reports, 6, 69
Non-secure Internet Protocol
    Router Network (NIPRNet),
    243
North Atlantic Treaty Organization
    (NATO)
  *See also* foreign nationals,
    organizations, and
    governments
  controlling office, 89
  information, 99
  member country abbreviations,
    **114**, 114
note (technical manual), 2, *3*
notices, 66, 67, 76
numerals, 301

## O

OPNAVINST 5510.161, 92
order of appearance of elements of
    a technical report, **24**
organization of technical reports,
    23
originator-controlled information,
    82, 115

# P

page format
  *See also* decimal numbering
    format; typographical
    progression format
  column format and line length,
    50
  description, 50
  examples. *See specific example*
  image areas and margins, 52,
    53, *54, 55*
  page numbers, 53, 66
  printers and paper and ink, 56
  typography, 50
paper and ink technical reports, 6,
  23
paragraphs, 26, 28, 34
patentable information, 10, 14, 16,
  88
performing organization
  distribution list, 221
  document control number, 71
  front cover and title page, 66, 72
  preface, 140
  relationship to author(s), 245
  report documentation page
    (Standard Form 298), 108
  technical report number, 70
person (grammar)
  *See also* tense (grammar); voice
    (grammar)
  abstract, 119
  acknowledgments, 141
  conclusions, 158
  executive summary, 147
  foreword, 136
  introduction, 156
  methods, assumptions, and
    procedures, 157
  preface, 140

person (grammar) (continued)
  recommendations, 158
  results and discussion, 157
  technical reports, 245
Portable Document Format (PDF)
  files, 15, 239
preface
  description, 137
  downgrade, declassify, and
    related markings, 94
  examples, *142, 143*
  export control notice, 92
  mention, 108
  order of appearance, 65
  relationship to foreword, 133
  technical report number, 70
  title and subtitle, 73
premature dissemination, 88
print technical reports, 6, 23
printers (occupation), xxiii
Privacy Act, 77, 78, 220
program element number, 107, 140
project and program managers, 10
project number, 107, 140
pronouns, 245, 246
proper names, derivatives of, 249
proprietary data/information, 82,
  87, 115, 157
publication date, 66, 72, 102
publishers, xxiii
publishing requirements, 8

# Q

QuestionPoint ("Ask a Librarian"),
  244
questions, use of, 247

# R

recommendations, 119, 145, 158

redundancy, 247
references, 156, 175, 188
references, list of
  comparison with bibliography, 192
  description, 161
  examples
    classified technical report, *178, 180*
    list, **181**
    unclassified technical report, *177, 179*
  number-identification system, 194
  order of appearance, 145
report documentation page (Standard Form 298)
  administrative information, 139
  description, 102
  examples, *112, 113*
  order of appearance, 65
  title and subtitle, 73
report number. *See* technical report number
Research and Development Descriptive Summaries (RDDS), 240
research, development, test, and evaluation (RDT&E), 8
restricted data and formerly restricted data, 83, 94, 96
results and discussion, 145, 157
review of technical reports, 10, 11, 12, 14

**S**

scientific and technical information (STI), xxiii, xxv, 8, 233

Scientific and Technical Information Network (STINET), 234, 243
scientific and technical reports. *See specific subject*
scientific and technical terms, 264
scientific names, 249, 301
scientific papers, 1
seals/emblems, 66, 101
SECNAVINST 5510.36, 98
Secret Internet Protocol Router Network (SIPRNet), 244
Sensitive But Unclassified (SBU) information, 77, 80
serial number, 66, 71, 223
series number, 66, 68
series of technical reports, 73
service names, 247
sexually biased language, 246
ship names, 247
slang terms, 248
Small Business Innovation Research Program, 84
software, 66, 84, 88
spacecraft, 300
Special Nuclear Material (SNM), 81
sponsoring/monitoring organization
  distribution list, 220
  front cover and title page, 72
  preface, 140
  report documentation page (Standard Form 298), 108
  technical report number, 70
Standard Form 298. *See* report documentation page (Standard Form 298)
standardization, xxiii–xxv, 4

State, Department of
  foreign country abbreviations,
    114
  Sensitive But Unclassified
    information, 77, 80
  U.S. Munitions List, 14, 91
*STI Handbook*
  description, xxvi
  reference, 8, 76, 85, 93
style, 248
style manuals, other, 305
subject terms, 116
subordination and format of
    technical reports
  *See also* decimal numbering
    format; page format;
    typographical progression
    format
  description, 25
subparagraphs, 31
subtitle of technical report, 66, 72,
    107, 118
summary, executive. *See* executive
    summary
supervisors, 10
symbols, 205, 265, **286**
symbols, abbreviations, and
    acronyms, list of
  description, 205
  examples, *210-213*
  order of appearance, 187

**T**

table of contents
  *See also* list(s) of figures and
    tables
  consistency with list(s) of
    figures and tables, 133
  description, 119

table of contents (continued)
  examples
    classified technical report,
      *127, 128, 130, 131*
    unclassified technical report,
      *126, 129*
  headings at the same level, 120,
    *122*
  minimum number of headings,
    121, *124*
  order of appearance, 65
  standalone element, 123
tables
  *See also* figures; visuals
  appendixes, 189
  classification markings, 26, *51*
  color, 49
  complementary role, 247
  components, *51*
  description, 49
  errata, 62
  examples. *See specific example*
  footnotes, 49, 61
  indexing, 218
  references, 176
  results and discussion, 157
  standalone elements, 132
  titles, 26
task number, 107, 140
Technical Area Tasks (TATs), 240
technical documents, other, 3
technical information centers
    (TICs). *See* libraries
technical manuals, 2 3
technical papers, 1
technical report number
  ANSI/NISO Z39.23-1997, 7
  back cover, 68
  description, 69
  foldouts, 41
  front cover and title page, 66

technical reports, scientific and.
    *See specific subject*
technology transfer, 14
tense (grammar)
    *See also* person (grammar);
        voice (grammar)
    abstract, 119
    acknowledgments, 141
    conclusions, 158
    executive summary, 147
    foreword, 136
    introduction, 156
    methods, assumptions, and
        procedures, 157
    preface, 140
    recommendations, 158
    results and discussion, 157
terminology, use of, 246
test and evaluation, 88
title of technical report
    *See also* subtitle of technical
        report
    back cover, 68
    description, 72, 73
    front cover and title page, 66
    keywords, 118
    report documentation page
        (Standard Form 298), 107
title page
    *See also* cover
    back
        description, 68
        examples, *108, 109*
        order of appearance, 65
    front
        authors, 74, 75
        DEA sensitive information,
            80
        description, 65
        distribution statement, 85
        DoD UCNI, 81

title page (continued)
    front (continued)
        downgrade, declassify, and
            related markings, 93
        examples, *104, 106*
        export control notice, 91
        foreign government and
            NATO information, 99
        mention, 108
        notice, 56, 57
        order of appearance, 65
        overall classification, 76
        restrictive markings on
            noncommercial data and
            software, 66, 84
        seals/emblems and logos, 101
        technical report number, 70
        unclassified controlled
            information, 77
        unclassified sensitive
            information, 81
        warning notices, 82
titles of persons, 251
tone, 245
trade names, 67, 157
trade secrets, 82
transmittal letters, 17
turnpage, 39
typographical progression format
    *See also* decimal numbering
        format; page format
    abstract, *120, 121*
    ANSI/NISO Z39.18-2005
        specification, xxvi
    appendixes
        description, 188
        examples, *190, 192, 193, 195*
    bibliography, 199, *204, 205*
    body of technical report, *159,
        163*
    description, 27

typographical progression format
(continued)
distribution list, 223, *224, 225*
errata, 61, *63*
examples. *See specific example*
executive summary, 147, *148,
150*
footnotes, 61
glossary, 210, *214, 215*
headings, 27, *29*
index, 218, *219, 220*
introduction, 156
list(s) of figures and tables, 133,
*134-136*
lists, 30, *32*
page format, 53
pages, 50
paragraphs, 28
references, list of, 166, *177, 178*
symbols, abbreviations, and
acronyms, list of, 208, *210,
211*
table of contents, 124, *126-128*
visuals, 37

## U

unclassified controlled information
abstract, 117
acknowledgments, 141
back cover, 68
description, 76
distribution, 89
executive summary, 147
foreword, 136
front cover and title page, 66
preface, 140

unclassified sensitive information,
77, 81
underlining, 246
*United States Government Printing
Office Style Manual,* 58, 75, 248
U.S. Munitions List, 14, 91

## V

vessels, 300
viewgraph compilations, 47, 247
visuals
*See also* figures; tables
color, 56
complementary nature of, xxvi
description, 34
examples. *See specific example*
executive summary, 56
facing pages, 41, *42, 43*
foldouts, 41, *44, 45*
numbering and titling, 37
orientation, *39, 40*
subfigures and subtables, 38,
*42, 43*
typography, 52
voice (grammar)
*See also* person (grammar);
tense (grammar)
active vs. passive, 246

## W

warning (technical manual), 2, *3*
warning notices
abbreviations, 26, 27, **115**
description, 82
front cover and title page, 66
work unit number, 107, 140

Figures - *Italic*; Tables - **Boldface**